ARE

SOULS

REAL?

A R E
SOULS
R E A L ?

Jerome W. Elbert, Ph.D.

Prometheus Books

59 John Glenn Drive
Amherst, New York 14228-2197

128.1

ELB

Published 2000 by Prometheus Books

Inquiries should be addressed to
Prometheus Books, 59 John Glenn Drive, Amherst, New York 14228–2197.
VOICE: 716–691–0133, ext. 207.
FAX: 716–564–2711.
WWW.PROMETHEUSBOOKS.COM

04 03 02 01 00   5 4 3 2 1

Library of Congress Cataloging-in-Publication Data

Elbert, Jerome W., 1942–
    Are souls real? / Jerome W. Elbert.
        p. cm.
    Includes bibliographical references and index.
    ISBN 1–57392–791–0
    1. Soul. 2. Religion and science. I. Title.

BD421 .E43   1999
128'.1—dc21   99–056822
                                                                  CIP

Printed in the United States of America on acid-free paper

To
Nancy Mottet Elbert,
my wife and friend,
for her kindness,
understanding,
and respect

CONTENTS

PREFACE

MANY PEOPLE BELIEVE THEIR SOULS are their most important possession. To them, only their souls have eternal value, and they devote their lives to caring for them. Many others are not so sure. Along with some theologians, they suspect that souls are fictitious entities. Unfortunately for the doubters, the local experts on souls are usually their pastors. Not surprisingly, spiritual counselors almost always defend beliefs in their favorite kinds of spirits, which usually include souls.

For anyone who is at all uncertain, *Are Souls Real?* offers an alternative to asking one's spiritual advisor about souls. Various experts, from biblical scholars to neuroscientists, have gathered information that allows soul beliefs to be judged more skeptically. This book brings these conclusions together, offering a new perspective on whether supernatural souls really exist.

In many ways, questions about the soul involve conflicts between religion and science. Even as we enter the twenty-first century, there is a danger that quarrels between religion and science will be resolved by ignoring science's answers. When I was a senior in a church-supported high school, I was told that it was permissible to believe in evolution as long as I also recognized that supernatural souls are an essential part of our nature. That was in Iowa, almost forty years ago. Now, for basically religious reasons, Kansas is dropping requirements for teaching evolution in science classes in *public* high schools. Clearly, some of the battles between science and religion are continuing. My own scientific training and study of issues related to religion lead me to believe that the battles are just beginning. Besides traditionally troublesome areas of biology, astronomy, and geology, quarrels will probably develop over topics in physics, psychology, medicine, and the law. Let's hope these quarrels do not lead to the exclusion of all these troublesome areas from Kansas's curriculum!

Without appealing to the idea of the soul, science is making tremendous progress in explaining our basic nature. *Consciousness* is an area of particularly rapid scientific progress. Its existence was long regarded as a reason for believing in supernatural souls. Now, consciousness seems to be yielding to scientific understanding. Ironically, the new understanding shows that *a supernatural soul lacks the properties needed to support consciousness.* This casts doubt on the idea that a soul, by itself, could support meaningful *personal immortality.* This argument also extends to other *spirits,* raising a problem for the idea of *angels,* and even the idea of *a personal God.*

For many years, I have been convinced that modern physics disagrees with traditional religious ideas about *free will,* which is supposedly one of the soul's powers. Many philosophers and scientists agree with me on this. If traditional ideas about free will are faulty, this has enormous consequences for religion and the law.

This book discusses issues from many different fields of learning. Since few people are acquainted with all these fields, I have defined special terms the first time they are used. There is also a glossary for technical or specialized terms that are used frequently, especially those mentioned a number of pages after their definition.

I have relied on many prominent experts whose ideas mesh, at least in some ways, with my own. In biblical scholarship, I made abundant use of books by Burton L. Mack and Robert W. Funk, both outstanding members of the Jesus Seminar. On the nature and physical bases of consciousness and emotions, I relied heavily on books by Daniel C. Dennett, the philosopher, and the following scientists: Francis Crick, Bernard J. Baars, Gerald M. Edelman, Antonio R. Damasio, and Joseph LeDoux. Concerning free will and questions of personal responsibility, I have been influenced and encouraged by the ideas of the philosopher Ted Honderich and, again, by Daniel C. Dennett and Francis Crick.

I would like to thank Paul Kurtz, Steven L. Mitchell, and all those at Prometheus Books who have supported this book's publication. I am grateful for research privileges granted by the Collins Library at the University of Puget Sound and the libraries of the Tacoma and Seattle branches of the University of Washington.

Special thanks to friends and family members who have read parts of the manuscript and have given their advice. Although they may not agree with some of the conclusions, they encouraged my early work and helped to shape this book. They include my wife, Nancy Mottet Elbert; John Graham; Carolyn Fox; Arthur Mottet Jr.; Arthur Mottet Sr.; Mary Louise Fox; and Gretchen Mottet. Many thanks to my friend and former colleague, Paul Sommers, for his reading of the manuscript and his comments. Although these people have generously given their time to this work, the entire responsibility for any errors or inaccuracies rests on me, the author.

I am grateful to Don W. Fawcett, M.D., Hersey Professor of Anatomy and Cell Biology, Emeritus, Harvard Medical School, for his permission to use the figure portraying the microscopic structure of skeletal muscles. Thanks also to Edward Arnold (Publishers) Ltd. for their permission to use the figure.

I would also like to thank my parents, June and Milferd Elbert, for their early encouragement of serious scholarship, and for all they have done for me over the years, especially during my childhood and education.

Jerome W. Elbert
Tacoma, Washington

Part I

RELIGION AND THE SOUL

1

CHALLENGING OUR IDEAS ABOUT THE SOUL

D O YOU HAVE A SUPERNATURAL soul? If you think you do, have you asked yourself *why* you think so? Most of us are guided more by tradition than by reason in forming our beliefs about such basic issues. Suppose, however, that there is no such thing as a soul. If the idea of a soul is mistaken, wouldn't you want to know? In this case, one's most precious possession is not a soul, but one's consciousness. What we believe about the soul and consciousness has an enormous impact on how we feel about life and our own nature. It also affects one's ideas about religion and the possibility of immortality.

I will argue that *our traditions give us an archaic and misleading view of the world and our nature.* Some of our ordinary ideas about the soul and what it is to be human appear to be mistaken. Recent results from many scholarly fields lead to ideas about human nature and our relation to the world that are radically different from the traditional views. These results will affect what we believe about such basic issues as *the nature of our consciousness, our freedom to choose right or wrong,* and even *what Jesus actually taught his followers.* The following chapters will lead you to a new understanding of your own nature, to freedom from guilt, and to reduced fear of death. The arguments and conclusions will tend to make you more tolerant, both toward yourself and toward others. You will obtain new perspectives on religion, crime, and punishment. The discussions will be based mainly on reasoning and scientific evidence rather than on tradition.

Everyone has a mental picture of what the world is and how it works. Such a picture, or *worldview*, helps a person guide his or her actions. A large part of our worldview comes from other people at home, school, work, and church, or from television, books, or other mass media. Our worldviews are strongly affected by our cultural traditions. In our culture, the Jewish and Christian religions have developed standard ideas that help to determine our worldviews. For Christians, one of these standard ideas is that the Bible's New Testament is an accurate report of the ministry of Jesus. Recent works of biblical scholars have concluded that *only part of the material in the New Testament reflects the actual life and teachings of Jesus*. These new findings are discussed in chapter 4.

Another standard idea in our culture concerns the relationship of mental processes to the material world. This relationship determines our attitudes toward the soul, consciousness, free will, moral responsibility, punishment, and nature. Many of the standard ideas were introduced into the cultural mainstream during the first five hundred years of the Christian Era by such theologians as St. Augustine. Some of these ideas are not in harmony with progress that has been made in a number of scientific fields, including the study of the human brain and mental processes.

In later chapters, *we will find that some of our standard ideas about mental processes are obsolete and mistaken*. I will describe new approaches to the issues of consciousness and free will, including *a model of the development of consciousness as a natural process*.

HOW SCIENCE OBTAINS
OBJECTIVE KNOWLEDGE

Why should we trust science when we deal with these important issues? *Much of the difference between a typical modern person's life and a life lived three hundred years ago is due to the successful application of scientific theories to our everyday life.* This progress has been possible because science can obtain relatively secure knowledge about the world.

Science is a systematic process of learning about the world by observing and testing its behavior. Scientists think of questions concerning phenomena that are to be investigated. They attempt to explain these phenomena by constructing a theoretical model or *hypothesis*. If the hypothesis is consistent with known facts it may deserve additional testing by experiments or observations. A hypothesis that has been successful in explaining phenomena and passing experimental tests, and which is more successful than competing hypotheses, may become part of the accepted body of theory in a scientific field. Of course, many hypotheses fail when they are tested and are rejected.

There are many variations in how scientific work is done. A hypothesis may be tested by a different person than the one who proposed it. Sometimes the results of different experiments are inconsistent with each other. Then it may be decided that certain experiments were flawed, or scientists may reserve judgment and wait for the results of even more experiments.

Even an accepted theory needs checking against new experimental results, since new results may overturn the theory. An interesting example from physics involves Newton's laws of motion. These laws, introduced in 1686, are extremely successful in describing motions of projectiles, planets, and other objects. They were used to build the impressive body of knowledge included in physics and parts of chemistry, astronomy, meteorology, and other fields of science and engineering. In 1905, however, Einstein proposed his *special theory of relativity*. Einstein's theory gave predictions that differed from those of Newton in certain unusual situations, such as when particles move at nearly the speed of light. Einstein's theory was found to be correct and Newton's theory was wrong in situations where the two theories made different predictions. This example shows that no scientific theory may be regarded as "holy," in spite of the superb reputation of the theory's founder or the previous successes of the theory.

Because science places primary importance on the results of experiments, the theories that are accepted stay in close agreement with reality *as we can detect and measure it*. The experimental method provides evidence to our senses. Properly trained people from any background or culture can verify and accept such evidence, even if they had different preconceptions of what the results would be. Thus, the results of science are *objective*. Scientific results that are obtained in Bombay, Bern, or Boston can be independently verified in Beijing, Berkeley, or Brussels. Because of this the same science exists throughout the world.

Scientists are often surprised by the unexpected results of experiments. This is because the experiments are adding to our knowledge about the world, not just confirming the prejudices of the scientists. The scientific process of obtaining knowledge produces well-established theories that can be trusted to describe the behavior of the real world.

WHEN RELIGION DISAGREES WITH SCIENCE

Although science is important in determining a modern person's worldview, one can ask how much weight it should be given in cases where religion and science disagree. To answer this, consider some historical examples. Quarrels between clerics and scientists have often ended in embarrassment for the religious side.

The ideas of Copernicus (1473–1543) that the earth is not the center of the solar system and that the earth turns on its axis were not accepted by Luther, Calvin, or the Roman Catholic Church. Luther discussed Copernicus's views on 4 June 1539, four years before Copernicus's major work was actually published. More than one version of Luther's *Table Talk* discussion of Copernicus exists. I will quote the more moderate version here.

> Mention was made of some new astrologer [*sic*] who would prove that the earth moves and not the heaven, sun, and moon, just as if someone moving in a vehicle or a ship were to think that he himself was at rest and that the earth and the trees were moving. [Luther responded as follows] "But this is the way it goes nowadays: anyone who wants to be clever should not be satisfied with the opinions of others. He has to produce something of his own, as this man does, who wants to turn the whole of astrology upside down. But even though astrology has been thrown into confusion, I, for my part, believe the sacred Scripture; for Joshua commanded the sun to stand still, not the earth."[1]

In opposing Copernicus, Luther based his argument on an Old Testament scripture, Josh. 10:12–14. This is a miracle story, set in the time of the Israelites' bloody invasion of Palestine. In this story, God answered a request by Joshua and extended the length of a day. By holding the sun still in the sky, God helped the Israelites carry out their wish of avenging themselves more fully against the Amorites, who had attempted to defend their homes and families. Luther's point was that if the sun's apparent movement were due to Earth's rotation, the Bible would have said that God stopped that rotation, rather than saying that the Sun's motion through the heavens had been stopped.

Calvin apparently referred to Copernicus in a sermon that was probably preached in 1556. Calvin admonished his listeners to avoid the error of

> those dreamers who have a spirit of bitterness and contradiction, who reprove everything and pervert the order of nature. We will see some who are so deranged, not only in religion but who in all things reveal their monstrous nature, that they will say that the sun does not move, and that it is the earth which shifts and turns. When we see such minds we must indeed confess that the devil possesses them, and that God sets them before us as mirrors, in order to keep us in his fear.[2]

The Roman Catholic Church did not actively oppose Copernicus initially. It did attack the Copernican ideas later, in connection with disputes with Galileo. Cardinal Bellarmine wrote a letter describing the outcome of the initial proceedings against Galileo in 1616. He stated,

> It is set forth that the doctrine attributed to Copernicus, that the earth moves around the sun, and that the sun is stationary in the center of the world and does not move from east to west, is contrary to the Holy Scriptures and therefore cannot be defended or held.[3]

In 1633 Galileo was convicted by the Catholic Church of continuing to support the Copernican theory. He was sentenced to a sort of comfortable house arrest and required to recite the seven penitential psalms once a week for three years. Then, quoting Arthur Koestler, "The recital of the penitential psalms was delegated, with ecclesiastical consent, to his daughter, Sister Marie Celeste, a Carmelite nun."[4] Books teaching that Earth moves were put on the Church's Index of Forbidden Books. This censorship finally ended in 1835. In 1979, 337 years after Galileo's death, the Church formally accepted Galileo's teachings and Pope John Paul II pardoned him.

In previous centuries many Christian authorities believed in a lifetime of Earth of about six thousand years, based on the generations and lifetimes mentioned in the Old Testament. Some religious fundamentalists still hold this view. This lifetime was made very specific in 1654 by James Ussher, the Archbishop of Armagh in Northern Ireland. He put the creation of Earth in the year 4004 B.C.E. Science, using methods based on the average lifetimes of certain radioactive atomic nuclei, gives about 4.5 billion years as the age of Earth.[5] The age of the universe is not determined precisely by present science, but is in the range of ten to twenty billion years.

The preceding examples do more than argue against accepting the Bible as literally true. *The point is that religion is fallible.* It does not have objective methods of acquiring the truth. In the areas where sacred scripture or religious traditions can be tested, they have sometimes been convincingly contradicted by science. It is too easy, with unfailing hindsight, to argue that the problems between scripture and science do not involve the real fundamentals of religion. *There was no advance warning by religion that the beliefs were not certain to be true in the cases where religion was found to be in error.* In fact, people were sometimes threatened with torture or burning at the stake for holding views that may now be dismissed as unimportant.

Imagine a world in which the results would have been different. In that imaginary world, the sacred scriptures came directly from God. We can reasonably assume that God would have been truthful and that the world would have been found to work exactly as implied by the scriptures. By the grace of God faulty translations or misinterpretations of the Bible would not have arisen. Any properly performed scientific experiment in that world would have shown that the Bible had been right all along. Science would have found, for example, that humanity had originated about six thousand years ago. This is what many people in our own world expected, but it has not turned out to be that way.

Religious faith is partly supported by the belief in miracles. Usually, miracles are supposed to have happened long ago, before there were modern methods of verifying such significant events. There are also some reports of modern miracles, but these *modern miracles are usually persuasive only to people who already belong to the particular religion that the miracles support.* Such forms of evidence as supernatural dreams or *visions* observed by individuals are so subjective that even if modern equipment is present, there is no way of verifying these occurrences.

According to the Acts of the Apostles in the New Testament, a persecutor of Christians named Saul of Tarsus was converted to Christianity after he experienced a vision while he was on the road to Damascus. Saul is reported to have seen a bright light and to have heard a loud voice. After the incident, he was blind for three days. Saul went on to become St. Paul, one of the most important missionaries of the Christian religion.

As will be seen in chapter 4, there is reason to believe that many of the supernatural occurrences reported in the New Testament are fictional. Even if the story of Paul's conversion is largely true, we may suspect that Paul was a victim of a hallucination, perhaps related to some disability. One author[6] has noted that Paul was "possessed of a 'thorn in the flesh, a messenger of Satan,' which has been interpreted as epilepsy."

Near the end of World War I, Adolf Hitler also experienced a dramatic spiritual conversion after a "supernatural" vision.[7] Hitler had suffered from a British mustard gas attack in 1918 and was hospitalized at Pasewalk, north of Berlin. While recovering from blindness, he suffered from depression and mental instability. During this time he experienced a vision, in which he said that he received a command from another world to save Germany. This vision convinced him that he was the agent chosen by a divine power to redeem his defeated nation. Morris Berman, in his book *Coming to Our Senses,* noted that Hitler's speaking ability was transformed by this incident from being relatively ineffective into being very powerful and charismatic.

Other murderers, of lesser fame than Hitler, have been inspired by what appeared *to them* to be supernatural visions or voices. Unless we are to believe *all* of these people, we are being sensible only if we are skeptical of attaching any special significance to *any* report of a supernatural vision, a voice from God, or less spectacular reports of religious or mystical experiences. Perhaps the person reporting the experience really did have it or, at least, seemed to remember the experience. On the other hand, we should not believe that such experiences really have a *supernatural origin or that they bring any validation of the message delivered in the experience.* In the irreverent words of Bertrand Russell, "From a scientific point of view, we can make no distinction between the man who eats little and sees heaven and the man who drinks much and sees snakes."[8]

In previous centuries, some theologians have given "proofs" of God's existence and other religious doctrines. Some of these were demolished as science described how biological evolution causes animals and plants to make beautiful adaptations to their environments. Some proofs were weakened when it was found that the solar system and the stars move by natural means, without the assistance of angels or the hand of God. *Not one of the so-called proofs has been accepted by most philosophers who are outside the religious group that presented the proof.* Consequently, it is fair to be skeptical of these religious proofs.

This section has shown that scriptures, miracles, traditions, mystical experiences, and theological arguments are all subject to skepticism. Consequently, *we have no compelling reason to grant religion higher authority than science in answering questions about the world.* If skepticism is appropriate concerning the literal truth of all religious doctrines, then science and philosophy are free to pursue all questions in spite of any previous doctrines.

THE SHIFTING BOUNDARY BETWEEN SCIENCE AND PHILOSOPHY

From what was argued above, all topics involving real or hypothetical phenomena are open to investigation by science, even if religion has classified the phenomena as supernatural or sacred. This includes such issues as the nature and origin of life. Even issues that are *outside the range of direct scientific investigation* may be discussed in light of the results of science. Topics that are usually thought to be issues of pure philosophy or religion may be affected by our scientific knowledge. For example, if life and human consciousness become understandable in scientific terms, some of our reasons for believing in a soul as a separate entity from the body will be greatly diminished. Under these circumstances the reasons for believing in the immortality of the human soul are greatly reduced as well.

Science has had a major impact on religion and philosophy in the last 450 years. *This process is far from finished.* The general public holds ideas about free will, spirits (souls, angels, devils, and God), and supernatural occurrences that will probably be regarded as unbelievable in the future.

As time passes, science will explain more and more about how our world works and how we live, think, and feel. Presumably the scientific explanations will not need to refer to supernatural processes. *In principle, however, science could find clear evidence for supernatural processes.* For example, certain decisions made by a person's brain *could* be found to be impossible to understand by the laws of science. If these results were found to make sense only in terms of, say, God's grace allowing a person to live a virtuous life, then there would be scientific evidence supporting supernatural processes. As other examples, video cameras could record miraculous apparitions, or certain outcomes of real-life situations could be shown to be more likely if people pray for them.

The examples given above show how science could detect supernatural phenomena if they actually play a role in our world. Any scientist who found such an effect would be involved in a giant controversy, but this scientist would be a celebrated hero if it turned out that he or she was not mistaken. There is plenty of motivation for detecting any supernatural effects if they exist. At present, of course, science finds no convincing evidence of supernatural phenomena.

NOTES

1. B. A. Gerrish, *Articles on Calvin and Calvinism*, ed. Richard C. Gamble, 14 vols. (New York: Garland Publishing, 1992), 12:6.

2. Ibid., p. 236.

3. Arthur Koestler, *The Sleepwalkers: A History of Man's Changing Vision of the Universe* (New York: Macmillan, 1959), p. 269.

4. Ibid., p. 500.

5. Francis Albarede, *Nuclear Methods of Dating*, ed. E. Roth and B. Poty (Dordrecht: Kluwer Academic Publishing, 1989), p. 52.

6. Homer W. Smith, *Man and His Gods* (New York: Grosset & Dunlap, 1952), p. 178.

7. Morris Berman, *Coming to Our Senses* (New York: Bantam Books, 1990), pp. 280–82.

8. Bertrand Russell, *Religion and Science* (New York: Oxford University Press, 1961), p. 188.

2

THE ANCIENT ORIGINS
OF THE IDEA OF THE SOUL

Gregory of Nyssa tells us Plato asserted that the intellectual substance which is called the soul is united to the body by a kind of spiritual contact; and this is understood in the sense in which a thing that moves or acts touches the thing that is moved or is passive. And hence Plato used to say, as the aforesaid Gregory relates, that man is not something that is composed of soul and body, but is a soul using a body, so that he is understood to be in a body in somewhat the same way as a sailor is in a ship.

—St. Thomas Aquinas, *On Spiritual Creatures*

THE IDEA OF THE SOUL

THROUGHOUT MUCH OF CHRISTIAN HISTORY, the *soul* has been thought to be the source of our most precious possessions: the ability to think, one's conscience, and free will. The soul is also regarded as that portion of a person that survives death. The word "soul" is supposed to refer to the part of ourselves that is not part of the physical world and is not based upon the physical world. Sometimes "spirit" has also been used in a religious sense to refer to part or all of our reputed supernatural aspects or abilities.

Supernatural entities, such as angels and souls, are believed to be outside the normal operation of physical laws. They are usually thought to be immaterial, in the sense of not being physical, and without definite weights or volumes. They are also often believed to be free of structural form, so they have no parts. They are often regarded as unchanging with time, so they are immortal. These supernatural essences are supposed to have a *spiritual* reality, but not a physical reality.

Somewhat inconsistently, although supernatural essences are not supposed to be part of the physical world, they often are described as producing physical effects. For example, they are often described as bright, or at least visible, or as producing sounds or having voices. So, they are imagined to have physical effects on certain occasions although they are not supposed to be part of the physical world.

It is important to point out a limitation to what we mean by "the supernatural." Some things that lack an accepted scientific explanation are not supernatural. For example, science does not have an adequate understanding of consciousness. If we think that consciousness will eventually be understood by science, then we do not think that consciousness is a supernatural occurrence. For something to be supernatural, it must be impossible for science to explain, not only now, but in the indefinite future.

It may be *impossible* to prove that science *will never* be able to explain something. Following this reasoning, we could never be sure that anything is really supernatural. Still, there are some things to discuss, since religious institutions have traditionally maintained that certain things *are* supernatural. It *does* seem reasonable to agree that a person's walking on water or rising into the sky will never be explained by science. As will be discussed in chapter 4, however, there is plenty of doubt that such New Testament stories actually happened.

Guardian angels and demons are examples of supernatural essences that are present in Greek mythology and Christianity.

Guardian spirits, angels, and devils were also present in *Zoroastrianism*, an ancient Persian religion. Most of us do not believe that demons cause illnesses, but this was different when the New Testament was written. In a biblical miracle story, Jesus exorcised a demon that had made a man mute, enabling the man to speak.

As the traditional source of such human capabilities as consciousness, thinking, and making moral decisions, the soul is nearer home than angels or devils. Since the brain and the soul are both supposed to be involved in thinking and choosing actions, there must be interactions between them. For example, suppose a mosquito bites your arm, you decide to kill it, and you slap your arm, killing the mosquito. In this example, your eyes, ears, and skin may all receive sensory information about the mosquito. Your nerves carry the information to the brain, which detects the mosquito, since the sensory information matches patterns that your brain has learned to associate with a mosquito.

We may believe that the soul is also involved in killing the mosquito. After the brain identifies a pattern as due to a mosquito, the soul may become conscious of the fact that a mosquito is present. The soul may form a plan to swat the mosquito. Since the plan involves killing, a check on the morality of the proposed action may be required. The morality check might conclude that there is nothing wrong with swatting a mosquito, according to Christian values. The soul may then decide to carry out the action. This is communicated back to the brain, which coordinates the swatting and sends signals to activate muscles in the arm.

In this imaginary process, there has been communication from the brain to the soul and from the soul to the brain. This implies that physical causes can have spiritual effects, and that spiritual causes can have physical effects. In particular, if the decision to swat the mosquito was really made by the soul and would not have occurred by means of the body alone, then the body took an action that would not have happened without the intervention of the soul. *As a result, the outcome was different from the outcome predicted by science, which would predict the actions of the body without considering the effect of the soul.*

From what was said above, certain physical effects could be observed, in principle, that have no *physical cause*, since they are produced by *spiritual causes*. If such spiritual processes are real, their effects could be detected by physical instruments. *So far, these effects have never been detected.* Science, of course, would not predict such effects, since they would violate physical laws, or at least they would overthrow science as we know it.

ORIGINS OF BELIEFS IN VARIOUS KINDS OF SOULS

Some have claimed that a belief in some kind of soul that can exist separately from the body is present in all known cultures.[1] It is very remarkable, then, to propose that the soul may not be a real entity, as we are doing here. If it is maintained that the belief in a soul is mistaken, then the fact that *most people believe in a soul* requires some explanation. Do not forget, however, that the soul, as proposed in many religions, would produce results that violate the laws of physics, as explained above. There is no convincing evidence of any such violations of physical laws.

Throughout almost all of human history, people's *imaginations* greatly exceeded their *understanding* of the world. Using the powerful human imagination, various groups formed beliefs that tended to be *as bizarre as could be allowed by their understanding of reality*. The parts of nature that were poorly understood or lacking explanation were "explained" by myths. Since our remote ancestors *did not know what is natural*, they were willing to accept explanations that we would consider unbelievable, fanciful, and excessively supernatural. It is difficult for a modern person to imagine the fundamental ignorance of primitive peoples. It is also difficult for us to appreciate the extent to which *this great ignorance created the possibility of accepting all sorts of supernatural beliefs*.

Until modern times, the processes of life were mystifying to people. A living animal or human appeared to contain something that is absent in a dead animal or person. The differences between the appearance of a body before and after death were sometimes very minor. This situation suggested that some very subtle thing left the body at death. This thing could be called a *life soul*. The life soul might have started out as just a name for *what leaves when life "goes."* In some cultures, life souls are attributed to humans, animals, and plants.

Besides life, the mental activities, emotions, and temperaments of people seemed very mysterious and needed explanation. Explanations in terms of brain function, glands, and hormones were not meaningful in ancient times. In order to explain the existence of these abilities and qualities in people, an entity was invented which generates these things. Since it explains the personal properties of the individual, we will call it the *ego soul*. It could be attributed to animals as well as humans.

Dreams and trances were also mystifying to our remote ancestors. Ancient people found it easy to believe that *these things occur in another real world*, rather than resulting from special modes of operation of the brain. This other world, purely imaginary, but thought to be real, can be called the *spirit world*. In dreams and trances, people could be reunited with their loved ones and ancestors who were long gone. It was easy to believe that dead people lived on as *surviving souls* in the spirit world. In this world, people are immortal. *As we know, the ideas of a spirit world and a surviving soul continued to be held even after they were no longer associated with experiences in dreams or trances.*

The ideas of a spirit world and surviving souls are ideas about supernatural entities. There is no physical evidence that proves their existence, and there is no convincing reason for science to believe that they exist.

There *is* reason to believe that beliefs about spirits and souls have been held for tens of thousands of years. The first Americans arrived at least ten thousand years ago. Australia was first populated at least forty thousand years ago. After becoming inhabited, the Americas and Australia were quite isolated from influences from other parts of the world. Nevertheless, beliefs in souls and human spirits (as well as other spirits) were widespread among the native Americans and Australians when European colonists arrived. This suggests that, even when the first people arrived in Australia and the Americas, they brought soul beliefs with them, or they already were inclined to believe in souls.

The dream world or spirit world imagined by ancient peoples was very different from the real world. Visions of people could come and go without normal movements. Dreams or trances could also *carry the experiencer* to distant locations. This suggested that a person has another kind of supernatural soul, called a *free soul*, that can leave the body during dreams or during a trance. Although free souls may seem bizarre to us, *there are folktales in many cultures that suggest old beliefs in free souls*.[2] Consider the following old Dutch story of what happened on a Saturday evening, when a suitor was courting a young woman.

> When they had been sitting for a while in a room the girl became so sleepy that the boy said, "Just lean on my shoulder." So she did and soon she fell asleep. Suddenly he saw a bumblebee creeping out of her mouth and flying away. He became worried and thought his girlfriend was a witch. He therefore took his handkerchief and spread it over her face. After she had been sleeping for twenty minutes the bee returned. The girl then became so short of breath that she got blue in the face, and the boy, afraid that she would suffocate, took the handkerchief off her face. Immediately the bee crept into her mouth, disappeared into her body, and she awoke.[3]

In this story, the free soul was depicted as a bumblebee. A very old story of this type appears in the eighth century *History of the Lombards*. In that story the sleeper was a Frankish king named Guntram and the free soul, as it usually was in central Europe, was a mouse. In Ireland the role of free soul was played by a butterfly, in Estonia it was a dung beetle. In Japan it was sometimes a bee, wasp, or dragonfly. In ancient Greece there were other kinds of stories in which a person fell into a trance and appeared dead, *while he was observed at the same time in another location*.[4] These stories supported a belief in a free soul. All of these may have been something like primitive *miracle stories*, perpetuating the belief in a free soul.

So far we have mentioned four soul types: *the life soul, the ego soul, the surviving soul*, and *the free soul*. These soul types are found in many cultures. A Swedish linguist, Ernst Arbman, and his students analyzed modern and ancient soul beliefs of native North Americans and of peoples in India, northern Europe, and northern Asia. Members of many of the studied societies believe in the free soul, which can be separated from the body, and the *body soul*, which endows the body with life and consciousness. The body soul includes the life soul and the ego soul. Other anthropologists have confirmed the conclusions of Arbman and his students.[5]

Among many peoples, including the ancient Greeks, the surviving soul was considered to be the continuation of the free soul.[6] In this picture, a living person's free soul is immortal and it becomes the surviving soul after death. A rather extreme form of soul belief is what might be called a *comprehensive immortal soul*. In a living person, the comprehensive immortal soul gives the body life, personality, emotions, mental abilities, personal identity, and consciousness. Except for life, these are all supposed to continue after the person dies.

A myth that includes the idea of a comprehensive surviving soul is very attractive. It may comfort a person who is approaching death by promising an almost lifelike existence after death. The idea of a surviving soul may console the bereaved person when a loved one dies. The idea implies that the loved one is not annihilated by death. An eventual reunion with the dead person seems possible. It may allow justice to be obtained in a future world, while justice is often absent in this world. Belief in a surviving soul may also benefit a society in wartime, because it seems like a lesser tragedy to die in battle if eternal rewards are expected.

Since the idea of the immortal soul has so many powerful attractions, the belief in an immortal soul improves the "evolutionary fitness" of a religion which includes this idea. That is, a religion is more attractive, obtains more converts, and grows if it supports a belief in an immortal soul. As a result, very successful religions tend to have adopted this idea. This may explain why the idea of an immortal soul is present in most world cultures, in spite of the lack of convincing evidence suggesting that there really is an immortal soul.

From a scientific point of view, some abilities explained by the various types of souls are becoming understandable as natural human abilities. These will be discussed later. These abilities arise from the interactions and properties of matter within the human body. As a result, some "explanations" provided by the different types of souls seem unnecessary. Normally, science drops ideas that are not supported by scientific evidence and are not needed to explain anything. In fact, science usually *does* ignore the idea of the soul.

There is an amusing idea suggested by this discussion of the soul's origins. The ideas for the different components of the soul *arise from ignorance* of how life, emotions, mental characteristics, consciousness, dreams, and trances arise in nature. In a technically advanced modern society, many of the items in the list have scientific explanations. In the long run, only those things that are not understood by science are apt to retain their supernatural explanations. An advanced society may reject all but one or two soul types, resulting in a belief in a relatively poor spiritual afterlife. In a very ignorant society, perhaps *none* of the items in the list will be understood. All the items may be explained by soul types, and they may all be included in the society's idea of an immortal soul. *This allows the members of a very ignorant society to anticipate a very rich afterlife.* Ignorance really is bliss!

THE HEBREW IDEAS OF THE SOUL

The Hebrews and the ancient Greeks contributed to our modern ideas of the soul. It is fair to say that nobody can prove anything concerning the nature of the soul and no scientific evidence supports the idea of the soul. The history of the ideas about the soul among the Hebrews and the Greeks suggests that, lacking evidence, groups may entertain many possibilities. Certain ideas became popular during certain time periods, however, and these ideas had effects that crossed the boundaries that are often imagined to exist between the religions of neighboring peoples.

Knowing the roots of words used for the soul in various cultures may allow us to reach back as far as possible into the past to find the original ideas about the soul. Especially for such an abstract idea as the soul, it is interesting to see what concrete thing or things were originally connected with the idea. The Hebrew soul-word *nefesh* is connected with breathing. This suggests that it originally referred to a life soul, since breathing is evidence of being alive. In classical Greek, the most important soul-word is *psyche*, related to the word *psychein*, which means "to blow" or "to breathe." Similarly, the Latin word for soul is *anima*, meaning "air," "breath," or "life." The related Latin word *spiritus* is also related to breathing.

Even more than the origin of words, the *use* of words in ancient literary works can show us what meanings were associated with such words as "the soul." From the ancient Hebrews, we can use the early books of the Old Testament to show how "soul" was used. The translated meaning of *nefesh* in hundreds of uses is "life" or "soul." *It was apparently present in humans and in other animals.* Although its root seems to be related to "breath," it was sometimes identified with the blood. Because blood was thought to hold some kind of life force even after death, it was forbidden to eat blood. *Nefesh* could also represent a person, and it was responsible for all the sensations and emotions of a person.

Another Hebrew soul-word, *ruah*, has a root meaning related to "blowing" or "wind." In the Old Testament, *ruah* sometimes means "wind" or "breath," but sometimes it can be translated as "spirit" or "soul," in the meanings of "life" or "consciousness." Unlike *nefesh*, it was never associated with the blood. It was often used to refer to the "spirit" of Yahweh entering a person, giving the person the wisdom, motivation, and strength to make prophetic pronouncements.[7]

Until long after the Jews returned from exile in Babylonia, the Hebrew soul was not expected to experience a rich and happy existence after death. It might continue as a shadowy entity in *Sheol*, the netherworld, but this was supposed to be a dull existence without clear consciousness. At that time, Sheol was not a place of reward or punishment. This view was shared with their neighbors, the Phoenicians, as well as with the more distant Babylonians, Greeks, and Romans.[8] The Egyptians and Persians, however, believed in a richer afterlife than the Jews. During these early times, the Hebrews did not believe in a resurrection followed by divine judgment, or in punishment or reward after death. Jews desired "to work out the life here on earth to its fullest possible development."[9]

Later, in difficult times of political oppression, in which justice on earth appeared to be absent, some Jews changed their minds about life after death. The changes may also have been motivated by the greater interest in the ultimate fate of the individual that occurred among people of the eastern Mediterranean in the last few centuries before the Christian Era. In changing their beliefs, the Jews adopted beliefs similar to Persian (Zoroastrian[10]) beliefs in the resurrection of the dead, final judgment, and reward and punishment following judgment. This will be discussed in more detail in the next chapter. In the adopted picture, eternal life was a reward of the just, and Sheol became a place of punishment for sinners.

For some time, there was disagreement between the *Sadducees*, who held the older beliefs, and the *Pharisees*, who accepted the Persian beliefs.[11] Joel B. Green has pointed out two Jewish passages from about the first or second century B.C.E. which illustrate the newer soul beliefs.[12] One, in the noncanonical Book of Enoch, declares that, "all souls are prepared for eternity, before the composition of the earth" (2 Enoch 23:5). The second, in the apocryphal Wisdom of Solomon (8:19–20), says, "As a child I was naturally gifted, and a good soul fell to my lot; or rather, being good, I entered an undefiled body."

The Persian beliefs were widely accepted by Jews by the start of the Christian Era. At that time, Jewish beliefs about an afterlife emphasized a resurrected body united with a soul, with a fully human existence. The dead were assumed to be in a situation like sleep until the resurrection. The beliefs about the afterlife were not primarily concerned with the survival of the soul without the body.

EARLY GREEK IDEAS ABOUT THE SOUL

Greek beliefs about the soul changed during the last few centuries before the Christian era. Evidence of the very early Greek ideas about the soul can be found in the uses of various soul-words in early writings such as those of Homer. "Homer's works" may have been written by more than one person during about the ninth century B.C.E.

The most important Greek soul-word is *psyche*, the root of the modern word "psychology." In Homer, the *psyche* leaves a person who swoons or dies. At death, a person's *psyche* left the body and went to *Hades*, the underworld. The underworld was a dark and unhappy place. Like Sheol, it was not regarded as a place of reward and punishment initially, although it later was thought to be such a place. Since the word *psyche* is associated with "breath," the *psyche* originally may have been a kind of life soul that later took on the roles of free soul and surviving soul.[13] Like the Hebrew *nefesh*, the life soul in Homer seems to have physical characteristics that are breathlike or bloodlike.[14]

The psychological traits of an individual, in Homer's time, were not part of the psyche. This includes the emotions, drives, and intellect that make up the part of the body soul that we have called the ego soul. For these characteristics, the Greeks had other soul-words such as *thymos* and *noos*. The *thymos* was the *source of emotions* such as fear, joy, revenge, anger, and grief. *Noos*, or *nous*, can be translated as *the mind* or as *an act of mind*, such as a thought or a purpose. Some animals possessed a *thymos*, but none had *noos*. *The* thymos *and* noos *were parts of the body soul, and they did not survive death.*[15]

The *psyche* was believed to be the only part of the soul that survived death. Since the abilities to think and experience emotions were contained within parts of the soul that did not survive death, the dead person barely continued to exist. The departed had a shadowy existence and were referred to as *shades*. As Bremmer says of the souls of the dead, "On the whole they are witless shades who lack precisely those qualities that make up an individual."[16] In this, the Greek souls in *Hades* resembled the Jewish souls in *Sheol* of about the same era.

From works written much later than Homer, around the fifth century B.C.E., it appears that *psyche* was used in two senses: as "courage" or "high spirit" or as the principle that makes the difference between a living body and a dead body. The *psyche* could be thought of as the natural animator, or biological "life" of the body.[17] To *Thales*, the founder of Greek philosophy who lived in the seventh and sixth centuries B.C.E., *an object that could move itself under its own control showed evidence of possessing a psyche.* Since a magnet apparently can move itself toward a piece of iron, Thales thought that a magnet must have a *psyche.*

Most Greek philosophers before Socrates tried to understand everything in terms of matter and its interactions. This is called *materialism.* (Note that this philosophical system *does not imply an excessive personal devotion to material or financial success.* Some people tend to confuse the respectable philosophical system with the reprehensible personal failing.) The materialists argued that the soul is made of relatively intangible substances such as air or fire. Democritus, who introduced the idea of atoms, proposed that the soul is made of very mobile spherical atoms. There was little to support such explanations of the soul. I maintain that the early Greek attempts to use science to explain life and mental processes were very premature, and that *the weakness of these early materialist theories allowed occult or spiritual explanations to be proposed by Socrates and Plato.*

THE SOUL ACCORDING TO SOCRATES AND PLATO

The three great teachers of ancient Greece were *Socrates* (c.470 B.C.E.–c.399 B.C.E.), *Plato* (428 B.C.E.–c.347 B.C.E.), and *Aristotle* (384–322 B.C.E.). They all made important contributions to our worldview. The lives of the three were linked; Socrates taught Plato and Plato taught Aristotle. Socrates was frequently involved in philosophical discussions, but, like Jesus, *he never wrote anything himself*. Socrates and Plato, in a dramatic shift away from the materialist explanations of the *psyche*, introduced ideas about an immortal and spiritual soul into their philosophies. *Their ideas have had a profound influence on popular modern ideas about the soul.*

The largest part of what we know about the philosophy of Socrates is found in Plato's writings, in which Socrates is represented as speaking for himself. Socrates' ideas, *as presented by Plato*, may not be the ideas of the real, historical Socrates. Plato's ideas changed during his career, but he continued to attribute his ideas to Socrates. We are probably dealing with the ancient custom of a student attributing his ideas to the school's founder. In this case, scholars have some information about Socrates from other authors. They also expect that Plato's earlier writings tend to be closer to the original teachings of Socrates.

Before Socrates, the *psyche* was the entity whose presence makes the difference between a living and a dead body. In the living it was thought to be the source of the body's movements. In the dead it had a very limited existence. It "bore little resemblance to what we call 'mind,' since it was not thought of as the seat of consciousness, will or feeling; it was merely a sort of shadow or phantom of the dead person, which could appear to the living in their dreams."[18]

Socrates greatly increased the importance of the *psyche* or soul by treating *it*, without the body, as the real person. He taught that taking care of one's soul is more important than caring for one's body or possessions. *This was a revolutionary idea to the Greeks of Socrates' time.* It was contrary to their basic approach to the world. About immortality, Plato has Socrates say, in *Apology* 40c, "Death is one of two things; either the dead man is nothing, and has no consciousness of anything at all, or it is, as people say, a change and a migration for the soul from this place here to another place."

Thus, Socrates was apparently uncertain about the immortality of the soul. Aristotle was also uncertain about the soul's immortality. Plato, especially in his later writings, seemed quite confident that humans have an immortal soul. He also believed that our souls exist before we are born. He argued that our souls contain great stores of recoverable knowledge from this previous existence. In these beliefs, Plato may have been influenced by *Orphism*, an early mystery cult, as well as by the ideas of *Pythagoras*, a mathematician, philosopher, and cult-founder.[19] Let's briefly look at those ideas.

Orphism was believed to be based on the writings of a legendary poet and musician named *Orpheus*. In a myth involving Greek gods, the Orphic cult "explained" how human nature is a combination of a soul of divine origin and a body of a much lower nature. Cult members tried to live according to the divine part of their nature. They purified themselves from the body's degrading influence by abstaining from wine, meat, and sex. After death, a person's soul was judged and sent to *Elysium* (heaven) or to hell, but that was not all. After this reward or punishment, the soul returned to earth in another body. The cycle was repeated a number of times, but the soul eventually returned to its divine source.[20]

Pythagoras (c.580 B.C.E.–c.500 B.C.E.) founded a cult in a Greek colony in southern Italy. His cult, *the Pythagorean brotherhood*, emphasized silence, obedience, lack of possessions, frequent self-examination, and abstinence from certain foods. Pythagoras believed that reality is ultimately mathematical in nature. Many of his ideas resembled those of Orphism. *He believed that souls were reincarnated into animals as well as humans*. Pythagoras claimed he had memories from some of his previous lives. It is interesting that it was at about the time of Pythagoras that the belief in reincarnation became part of the Hindu religion.[21]

In his dialogue *Phaedo*, Plato attempted to prove that the soul is immortal on the grounds that it is impossible for the soul or *psyche* to die. He argued that "life" belongs to the *psyche* as an *essential property*. That is, a *psyche*, by its nature, is alive. For it to change into something that is not alive is contrary to its nature. Thus, Plato argued that a *psyche* must never die and is immortal. Recently, however, one philosopher pointed out a fatal flaw in this argument. An essential property of a thing is something that the thing must continue to have *if it continues to exist*. Death may involve the end of the *psyche's* existence, resulting in the loss of awareness.[22] Another possibility, of course, is that *there is no immaterial entity like the* psyche *that gives life to the body.*

Plato gave another argument for the immortality of the soul in *Phaedo*. He argued that the soul does not have parts. As an entity without parts, it can't break up or fall apart, so it never ceases to exist. Later, however, Plato caused a major problem with this argument when he taught that the soul consists of three parts: *reason, spirit* or *passion*, and *appetite*. In chapter 10, I will argue that a person needs billions of functioning parts (neurons in the brain) in order to be conscious. In this view, ironically, enormous numbers of parts are a necessary requirement for consciousness to be present.

In Plato's *Republic* there is a discussion of how to train the elite young people in an ideal society. Plato says that if the young believe in an immortal soul it is not good to describe *Hades* as a terrifying place. In that case, "Do you think anyone will be fearless, and choose death in his battles rather than defeat and slavery?" He goes on to advocate censorship of the fables taught to the youth, favoring myths that paint the afterlife as a pleasant existence. By this indoctrination he hoped to produce young men who would be fearless in battle.

Later in his *Republic*, Plato tells a story that he must have approved. It is about a man named Er, who was killed in battle and saw the place where souls were judged. By special treatment, the judges allowed Er to return to earth to inform people of what he had seen. Near some judges were two side-by-side openings going up to heaven and two other openings going down to the underworld. Souls could be sent up or down after judgment. They were rewarded or punished for a thousand years. Since the story assumed reincarnation, the *second doors* from heaven and the underworld were used by souls *on their way back* to obtain new bodies on earth. The souls had some choice of which bodies they occupied, and could become humans or animals. Afterward, the returning souls drank from a magic river that made them forget their previous experiences. Then they passed on to earth.

Let's consider some characteristics of the soul according to

Plato. Unlike the soul imagined by some previous Greek philosophers, Plato's soul is immaterial and immortal. As an immaterial
essence, it exists in a different realm than the world of our sensory
experience. Plato's *psyche* produces all mental activities in life, and
preserves these mental capabilities after death. To Plato, the soul is
the source of all movements of the body, and it is the natural
master of its subject, the body. Because it is the natural master and
the only immortal part of a human, the soul is superior to the body
and the most important part of a person.

Plato regarded the soul as the part of human nature that is
closest to the divine. Learning to take care of one's soul and to control the degrading influences of one's body is therefore a way of
coming to resemble God. Plato believed that the human soul, by its
godlike faculty of reason, is a creature "not of earth but of heaven."[23]
Thus, the soul was not a natural and complementary companion of
the body in human nature, but was more naturally an independent
spiritual being. Plato spoke of the body imprisoning the soul.
Because of Plato's great influence on later thought, he was probably
an important source of the otherworldly views of Christianity in its
first thousand years. We are still under Plato's influence.

EARLY CHRISTIAN SOUL BELIEFS

Some of the early followers of Jesus inherited Jewish ideas about
the soul. They probably believed the soul gives life to the body, but
is capable of only a weak existence without a living body. The early
Christians also inherited the belief in a future resurrection and
judgment. The hope for immortality was based on the reunion of
body and soul at the resurrection. In this picture, the body and the
soul are intimately connected and there is little reason to imagine a
bodiless soul enjoying heaven before the resurrection.

Some uses of the word *psyche* in the Greek of the New Testament show that some very early Christians shifted their ideas of the
soul from the Jewish ideas described above toward the Greek
beliefs of Plato and others. In Matt. 10:28 Jesus says, "And do not

be afraid of those who kill the body but cannot kill the soul; but rather be afraid of him who is able to destroy both the body and soul in hell." In Rev. 6:9 and 20:4 there are references to seeing the immortal souls of Christian martyrs. In these passages, the soul is treated as if it can be alive and functioning after death.[24] In Rev. 6:9 the souls are able to be seen, to cry out, and to wear white robes.

Most likely, there were always great variations in what individual Christians believed about the soul. Certainly, early theologians disagreed on soul beliefs. Tertullian (c.160–c.220) held that the soul is physical and develops along with the body. Origen (c.185–254), who had studied Plato's ideas, taught that the soul has a spiritual nature, that it exists before the body, and that it is immortal. Jerome (c.345–420) argued that the soul is created at the time of conception.[25]

Augustine, bishop of Hippo (354–430), was one of the most influential theologians in Christian history. For about eight hundred years from his lifetime to well into the Middle Ages, his views dominated Christian ideas about the soul. His thinking was strongly affected by Neoplatonism, a pagan, Jewish, and Christian philosophical movement built primarily on Plato's ideas, with lesser influences by Aristotle, Pythagoras, and others.

According to Augustine, a human being is an immortal, spiritual soul *using* a mortal body. For Augustine, the soul is very different from the body and far superior to it, so that it would not seem consistent that the body could affect the soul. Thus, it was difficult for Augustine to account for our ability to acquire knowledge by using our senses. *Significantly, Augustine introduced the traditional Christian idea of the will.* His teachings form a major link in a chain of soul beliefs reaching from Plato to present-day Christians.

AQUINAS AND ARISTOTLE

In western Europe, a major change in soul beliefs occurred when Islamic scholars brought Aristotle's works to Europe in the thirteenth century. Previously, medieval Europe had little knowledge of Aristotle's writings, except for his work dealing with logic.

Compared with Plato, his teacher, Aristotle tended to describe humans in natural rather than supernatural terms. He trusted the senses as a source of knowledge about the world, writing that "There is nothing in the intellect that was not first in the senses." While Plato had argued that we have a dual nature consisting of a spiritual soul and a body, Aristotle described the soul as the "form" of the matter in the body. For Aristotle, body and soul are more of a unity, just as "the wax and the shape given to it by the stamp are one."

Thomas Aquinas (c.1225–1274) built a grand synthesis of Aristotle's philosophy and Christian theology. Born in southern Italy, he studied at Monte Cassino, Naples, Paris, and Cologne, and worked in Paris, Rome, and Naples. Among other subjects, Aquinas dealt with the soul.

Aquinas argued for three levels of faculties in the human soul.[26] At the lowest level are the "vegetative" faculties of nutrition, growth, and reproduction. Animals and plants also have these faculties.

Next come "sensitive" faculties, held by animals and humans. These include the "exterior senses" of sight, hearing, smell, taste, and touch, along with four "interior senses." The latter include the "common sense," memory, imagination, and the "estimative power." This power allows a sheep to recognize a wolf as an enemy and allows a bird to realize that straw is useful for building a nest. The powers to move oneself and to have appetites and emotions are also parts of the sensitive faculties of the soul.

At the highest level, humans, but not other animals, possess the "rational faculties." These are the intellect and the will. The intellect includes such abilities as comprehending abstract ideas, and making abstractions and judgments. The will seeks the good. Morality involves using the intellect to determine what is good, and using the will to try to attain the good.

While Augustine regarded the will as greater than the intellect, Aquinas regarded the intellect as greater than the will. Augustine's soul beliefs easily led to a belief in the immortality of the soul, but made it difficult to explain how the body and soul produce a unified person. Aquinas's ideas of the soul led more naturally to unity of the person, but made it more difficult to explain the soul's immortality. His views also suggested that meaningful immortality would be attained only after souls are reunited with resurrected bodies.

Aquinas's teachings about the soul were debated fiercely for about a generation. Two archbishops of Canterbury attacked Aquinas's views. Aquinas's religious order, the Dominicans, supported his views. The Franciscans, a rival religious order, supported more traditional ideas based on Augustine and Plato, while opposing Aquinas's views. In 1323, however, Aquinas was declared a saint, and the criticisms subsided. Since then, his ideas have been highly esteemed by the Roman Catholic Church. In 1950 Pope Pius XII declared that Aquinas's philosophy is the surest guide to Roman Catholic doctrine. For everyday soul beliefs, however, Aquinas did not have the last word.

DESCARTES AND MODERN SOUL BELIEFS

René Descartes (1596–1650), a mathematician and philosopher, was one of the most influential contributors to modern ideas about the soul. His attempts to understand our basic nature, combining rational analysis with introspection, supported and reinforced traditional soul beliefs. By persuasively arguing that human nature consists of an immaterial soul and a physical body, Descartes provided intellectual respectability to soul beliefs that were already held as a matter of religious faith. Although his arguments have lost their appeal to most modern philosophers and scientists, his ideas survive in everyday thought and speech.

Born in France, Descartes was educated in classical studies, mathematics, and philosophy in a Jesuit school. Next, he studied law at the University of Poitiers, graduating in 1616. Two years later, he enlisted in a Dutch army led by Maurice of Nassau. He did not make his fame as a lawyer or a soldier, but his way of life gave him time to think. On November 10, 1619, alone in a "stove-heated room" in southern Germany, he began to "understand the foundations of a marvelous science" and was filled with great enthusiasm for a grand intellectual project.[27]

He decided to examine all of his beliefs, and to reject any which seemed even slightly doubtful. By doing this, he hoped to find at least one belief that seemed completely clear and certain. He would then accept this belief as true, and see what else he could conclude based on this undeniable belief.

Using his systematic method of doubting, Descartes realized that he had learned most of what he knew by using his senses. His method forced him to admit that he might have been mistaken, dreaming, or hallucinating while learning, so he concluded that he should not trust *anything* he had learned by sensory means. Consequently, he could not be sure that anything exists in the external world, or even that he possessed a body!

In the end, he found he could be certain of something. He realized that, while trying to show that everything was uncertain, it still must have been true that *he, who was thinking, was something.* Since he could find no reason to doubt it, he decided that it was true that, "I am thinking, therefore I exist." He accepted this as the first principle of his new philosophy.

From this undeniable fact, and the possibility of doubting that he possessed a body, he concluded that he was basically *a conscious being.* What else does he conclude about himself? In his Sixth Meditation, in his *Meditations on First Philosophy*, he writes:

> Now I know that I exist, and at the same time I observe absolutely nothing else as belonging to my nature or essence except the mere fact that I am a conscious being; and just from this I can validly infer that my essence consists simply in the fact that I am a conscious being. It is indeed possible (or rather, as I shall say later on, it is certain) that I have a body closely bound up with myself; but at the same time I have, on the one hand, a clear and distinct idea of myself taken simply as a conscious, not an extended, being; and on the other hand, a distinct idea of body, taken simply as an extended, not a conscious being; so it is certain that I am really distinct from my body, and could exist without it.[28]

By this reasoning, Descartes claimed that human nature has two parts. This belief in a dual human nature is sometimes called *Cartesian dualism*. One part of our nature, the body, has physical properties, such as occupying space (that is, it's an "extended" being). The other part, the mind or soul, is conscious and is the true self. (Descartes treats the mind and soul as interchangeable.) The soul, he maintained, does not have any physical properties and could exist without a body.

As an aside about "Cartesian dualism," I should note that soul/body dualism was already widespread in Christianity during the century *before Descartes*. The Catholic Church and the Protestant reformer John Calvin maintained that the dead enjoy a conscious relationship with God even before the general resurrection.[29] This implies that souls exist independently of the body, and are conscious, between death and resurrection. The Catholic position was affirmed at the Fifth Lateran Council in 1513, and Calvin wrote about this in *Psychopannychia* in 1542. On the other hand, "Martin Luther and some of the Radical Reformers argued that the soul either dies with the body or 'sleeps' until the general resurrection. . . . Yet even those who argued for the death of the soul were obviously presupposing a dualistic conception of the person."[30]

Descartes *actually did not prove that the soul can exist without the body*. He was certain that his basic nature was that of a conscious being, and that he had a mind and a body that had entirely distinct properties. *But this merely suggested that the mind and the body might be independent entities*. Instead, Descartes may have identified different features of *the same entity*, viewed from radically different perspectives. If the mind and the body really are different features of the same entity, then *the mind would not be expected to exist without the body*! Thus, Descartes did not prove that the mind is independent of the body. Later on, I will argue that "mind" is a name we give to a group of capabilities of the human body that are primarily the result of physical processes within the brain.

Descartes's leap to conclude that the soul could exist without the body was partly a result of his background and education as a Roman Catholic. He appears to have been sincerely religious. In his

Private Thoughts, he wrote, "The Lord has made three marvels: things out of nothing; free will; and the Man who is God." His dualism, and his emphasis on free will and the soul's importance, put him in the tradition of Plato and Augustine, but less in the tradition of Aristotle and Aquinas.

Besides piety, Descartes had down-to-earth reasons why his speculations had to lead to what the Church regarded as the "right answers." He related how he delayed publishing a manuscript because he feared that it might be judged to be heretical. He thought it was free of heresy, but, without naming names, he noted that *he had also thought there was nothing wrong with Copernicus's theory*, but the Church had condemned Galileo for teaching it.[31]

Descartes emphasized the unity of body and soul, saying that the soul is "not present as a pilot is present in a ship; I am most tightly bound to it, and as it were mixed up with it, so that I and it form a unit."[32] Because of his dualism, this unity was not easily explained. If the mind and the body have entirely different natures, they ought to operate independently. In fact, there are interactions between them that go in both directions. The mind becomes aware of pain when the body is injured. The body acts after the mind has made a decision. How this works was left as a mystery, perhaps best explained by saying, "For God, all things are possible."

Descartes changed the way we think about ourselves by highlighting *consciousness* as the most important aspect of mind. His writings linked the terms "soul" and "mind," since both refer to the entity that supports consciousness. By creating his form of dualism, he produced a worldview that was reasonably satisfactory to the Church and, at that time, to the emerging sciences.

Even today, many people's beliefs are influenced by Descartes, or at least it can be said that they are remarkably consistent with his views. Many believe there is something mysterious about the mind, and that *matter, by itself, is not capable of consciousness*. These beliefs follow from Cartesian dualism. Although dualism implies that mental processes could occur without any physical activity, modern imaging devices show that purely mental processes do involve physical activity within the brain.

As a religious belief, many believe the soul is immortal. Cartesian dualism supports this, since the soul, being independent of matter, is supposed to be able to survive without the body.

Descartes held that animals (except humans) are not conscious, because they lack souls. Some scientific experimenters believed him and participated in cruel experiments on live animals, believing that the animals' cries were purely instinctive, not due to conscious pain.

Recently, scientists have often avoided using the word "mind" because of its Cartesian connections with the supernatural soul. In later chapters, I will use "mind" without meaning to imply that it is an entity that can exist without a body.

Before Descartes, the soul was primarily a religious and theological concept that was weakly linked with everyday life. Since Descartes, the *mind* has taken over many of the soul's mysterious characteristics. This injects an element of mystery (unnecessarily, in my opinion) into our feelings about consciousness.

SOME POPULAR AND QUESTIONABLE SOUL BELIEFS

Like Plato, Augustine, and Descartes, many Christians believe that humans have a profoundly dual nature, involving some linkage between a body and a soul. The Christian soul is often thought to support all mental processes such as *consciousness* and *reasoning*. It is also thought to be immortal and immaterial, consistent with Plato's ideas. Unlike the Jewish soul of Jesus' time, the Christian soul is often thought to allow a rich afterlife before it is reunited with the resurrected body. Unlike Plato's idea of the soul, the Christian soul is *eternal only in the future direction*. That is, it did not exist during the eternity of past times, but was brought into existence by God when it was needed to combine with a body to form a human being. The Christian soul, unlike Plato's, is not involved in reincarnation.

Some people believe that life itself is bestowed on a body by the soul. Clearly, in Christian thought, death involves the departure of the soul from the body. In a common Christian belief, animals do not have immortal souls. For animals an immortal soul is *not* believed to be needed for them to be alive. The belief in the lack of immortal souls in animals is an important reason why people often deny that consciousness and mental activities exist in animals other than humans.

Another important feature of the Christian soul is the ability to make moral choices by means of *the will*. Often, the will is thought to allow one to make choices that would have been made *differently if left to the brain alone*. The brain, if it operates according to scientific laws, could have produced only *one outcome* of the decision-making process. The will is supposed to allow humans to choose between two possible outcomes, *with either choice being possible*. Since, in a particular situation, science would predict *only one outcome of the brain's decision-making process, the will's fundamental ability to choose* implies that an outcome could be selected that does not follow the laws of science. Such a will clearly bestows a *supernatural* ability to override the laws of nature.

Other abilities commonly associated with the Christian soul are the ability to *reason* and the presence of a *conscience*. In this view, reasoning is beyond the capability of the brain by itself, and takes place with at least the assistance of the soul. The conscience is supposed to be a special ability to have correct inner knowledge on questions of good and evil. Since this ability is assumed to be more than the effect of an education in morality and one's ability to reason, it has been regarded by many as a supernatural ability.

It is very interesting to note that the supernatural functions of *consciousness, conscience, reason,* and *free will* are all features of the soul beliefs of the ancient Persian Zoroastrians as well. Thus these ideas are probably more than 2,500 years old. Christianity may have obtained these ideas independently or from the Zoroastrians, possibly through the influence of the Persians on the Greeks and the Jews. As will be noted in chapter 3, the Zoroastrian soul also included a guardian spirit. Those Christians who believe in a guardian angel believe that it is a separate spiritual entity, not part of the human soul.

If one believes in biological evolution as well as in the Christian soul, one may be led to imagine the following scenario of humanity's origin. Our soulless ancestors gradually developed more advanced brains until a point was reached when God said, "This is good enough," and infused immortal souls into the first humans' bodies. If these souls really endowed our ancestors with many new abilities, *then the first true humans were far superior to their immediate, soulless ancestors.* On the other hand, if humans evolved from animal ancestors by purely natural processes, there may never have been a sharp dividing line between the "first humans" and their parents.

NOTES

1. *Encarta 96 Encyclopedia*, s.v. "Soul," by John A. Saliba.
2. Jan Bremmer, *The Early Greek Concept of the Soul* (Princeton: Princeton University Press, 1983), pp. 132–35.
3. Ibid., p. 134.
4. Ibid., pp. 25–27.
5. Ibid., pp. 9, 10.
6. Ibid., p. 73.
7. R. B. Onians, *The Origins of European Thought About the Body, the Mind, the Soul, the World, Time, and Fate: New Interpretations of Greek, Roman and Kindred Evidence: Also of Some Basic Jewish and Christian Beliefs* (Cambridge: Cambridge University Press, 1951), p. 483.
8. K. Kohler, *Jewish Theology: Systematically and Historically Considered* (New York: Macmillan, 1923), p. 279.
9. Ibid., p. 281.
10. The Zoroastrian religion was founded by Zoroaster, an ancient Persian prophet. Its followers practiced a form of monotheism that incorporated dualistic concepts such as a global conflict between good and evil. It was adopted by some of the kings of the Persian Empire, but barely survives today among small groups in India and Iran. It is of special interest here because it appears to be a source of many Jewish and Christian supernatural beliefs. It is discussed more fully in chapter 3.
11. Kohler, *Jewish Theology*, p. 283. Near the beginning of the Chris-

tian Era, there were two schools of especially strict practitioners in Judaism. The Pharisees were essentially an upper-class religious sect, rarely involved in political turmoil, except when it involved religious issues. They were strict adherents to religious *traditions*. They were often opposed to the Sadducees, who were very strict supporters of the *written law*, as opposed to oral traditions. The Pharisees were more popular with the people, while the Sadducees tended to represent the highest classes, consisting of aristocrats and priests. The Sadducees were deeply involved in secular and political affairs. The Pharisees believed in the immortality of the soul, angels, demons, and a final judgment, while the Sadducees rejected such beliefs.

12. Joel B. Green, *Whatever Happened to the Soul? Scientific and Theological Portraits of Human Nature*, ed. Warren S. Brown, Nancey Murphy, and H. Newton Malony (Minneapolis: Fortress Press, 1998), p. 161.

13. Bremmer, *The Early Greek Concept of the Soul*, p. 24.

14. David B. Claus, *Toward the Soul: An Inquiry into the Meaning of Psyche Before Plato* (New Haven: Yale University Press, 1981), p. 61.

15. Bremmer, *The Early Greek Concept of the Soul*, pp. 54–57, 74–76, 126, 127.

16. Ibid., p. 124.

17. Claus, *Toward the Soul*, p. 101.

18. Sabina Lovibond, *Psychology*, ed. Stephen Everson (Cambridge: Cambridge University Press, 1991), p. 36.

19. Much of Plato's discussion of the soul is contained in *Phaedo*, known in ancient times as *On the Soul*. It is presented as a drama dealing with Socrates' last hours in a jail in Athens. The Pythagorean influences in this source are discussed in the introduction to Plato's *Phaedo in Plato: Complete Works*, ed. John M. Cooper, assoc. editor D. S. Hutchinson (Indianapolis/Cambridge: Hackett Publishing Company, 1997), p. 49. The possible interaction of Plato with the Pythagorean community in southern Italy is discussed on p. viii of the same volume. Peter Gorman, in *Pythagoras: A Life* (London: Routledge and Kegan Paul, 1979), notes that Plato borrowed a Pythagorean idea of three parts to the soul and used it for his own political arguments. Plato also borrowed the Pythagorean doctrines of reincarnation and recollections from previous lives, according to Gorman. These are mentioned on page 132 of *Pythagoras*. In *Cratylus* 400c, Plato refers to followers of Orpheus as the source of the idea that the body is a prison of the soul. There may also have been indirect Orphic influence on Plato through the Pythagorean influence. Orphic influences

on the Pythagoreans are discussed in *Pythagoras* on pages 89–91. That book also points out numerous similarities between the Orphic and Pythagorean beliefs.

Plato's conservative respect for the beliefs typical of Orpheus and Pythagoras is suggested in his *Letters* VII, 335a, in which he states: "And we must always firmly believe the sacred and ancient words declaring to us that the soul is immortal, and when it has separated from the body will go before its judges and pay the utmost penalties." This can be found, for example, on page 1654 of *Plato: Complete Works*.

20. Orphism is described in more detail in W. K. C. Guthrie's *Orpheus and Greek Religion: A Study of the Orphic Movement* (New York: W. W. Norton & Co., 1966).

21. See Peter Gorham's *Pythagoras: A Life* (cited above in note 19) for more about Pythagoras.

22. Lovibond, *Psychology*, p. 45.

23. Ibid., p. 55.

24. *New Catholic Encyclopedia* (1967), s.v. "Soul (In the Bible)," by W. E. Lynch.

25. Murphy, *Whatever Happened to the Soul?* p. 4.

26. Ibid., pp. 5, 6.

27. René Descartes, *Descartes: Philosophical Writings*, trans. Elizabeth Anscombe and Peter Thomas Geach (Indianapolis: Bobbs-Merrill Co., 1971).

28. Ibid., pp. 114, 115.

29. Murphy, *Whatever Happened to the Soul?* p. 23.

30. Ibid., p. 19.

31. Descartes, *Descartes: Philosophical Writings*, p. 45.

32. Ibid., p. 117.

3

THE PAGAN PARALLELS
OF SOME CHRISTIAN BELIEFS

None cometh from thence
That he may tell us how they fare.
Lo, no man taketh his goods with him.
Yea, none returneth again that is gone thither.
    —*The Song of the Harp Player*, Egyptian, c.2100 B.C.E.

The soul and various spirits, such as *angels* and *Satan*, are part of the Christian heritage. Jesus performs many miracles in the New Testament. Many Christians believe these supernatural phenomena are as real as the world in which we live. On the other hand, many of the same Christians believe that all of the gods, spirits, and supernatural phenomena of other religions are false. Those people tend to draw a clear distinction between all other religions and the particular religion that they inherited or adopted.

There are many reasons for doubting that Christian beliefs are uniquely inspired truths that are unlike the beliefs found in the pagan religions. In fact, it appears that many Christian beliefs about the supernatural are related to, or drawn from, the earlier mythological traditions of such pagan peoples as the Babylonians, Egyptians, Persians, and Greeks.

OLD TESTAMENT MYTHS
SHARED WITH OTHER PEOPLES

Before we discuss pagan ideas that are also present within the New Testament, let's consider some examples from the Old Testament. The Hebrew people were located between, and not far from, the two earliest civilizations: Mesopotamia and Egypt. Not surprisingly, the Old Testament contains stories that are similar to myths found in older Egyptian and Mesopotamian writings. It is especially reasonable that there are some connections between the ancient Hebrew beliefs and those of Mesopotamia if we accept the tradition that the Hebrew patriarch Abraham migrated to Palestine from Mesopotamia.

There are similarities between the story of Adam and Eve and parts of mankind's earliest great epic, the Babylonian epic of *Gilgamesh*. In the Gilgamesh epic there is a reference to Eden as the most fertile country in the world. In Eden, a hero named *Engidu* or *Enkidu* was formed out of clay by the goddess *Aruru*. Enkidu lost his immortality by being seduced by a woman. However, Enkidu was not the first man in these myths. That was *Adapa* or *Adamu*.[1] In another part of the epic, Gilgamesh searches for a plant that will give him immortality. He finds it, but loses it to a serpent, and resigns himself to being mortal.

There is a great similarity between the story of Noah and the Flood and another story in the epic of Gilgamesh. In the epic, the story was told to Gilgamesh by *Utnapishtim*, who takes the place of Noah. A god told Utnapishtim that a flood was coming and that he should build a ship for his family and animals. The ship was built, Utnapishtim survived the flood, and the ship came to rest on the top of a mountain after seven days. The story of the Tower of Babel is also a biblical story that is present in Mesopotamian mythology.[2]

A story about Moses as an infant is another example of a myth shared by different groups. Moses was born in secret and hidden in a basket in bulrushes to hide him from the pharaoh who wanted to kill him. In an Egyptian myth, the goddess *Isis* gave birth to the god *Horus* in the Nile Delta. Isis hid the infant in a basket of rushes so that he would not be killed by a jealous god. *Sargon*, an important early ruler of Mesopotamia, was also said to have been born secretly. According to legend, Sargon's mother was a virgin. He was placed in a basket of rushes and set adrift in the Euphrates River. Moses, Horus, and Sargon were all rescued, of course, and each played an important real or mythical role as an adult.

EGYPTIAN BELIEFS ABOUT AN AFTERLIFE AND MORALITY

Our impressions of ancient Egyptian beliefs about the supernatural often come from *images* of mummies and various gods depicted as bizarre combinations of human and animal body parts. If we look at the Egyptian *beliefs* about the afterlife, however, some of their ideas do not seem so strange. They believed that humans can be resurrected and have eternal life. The resurrected person, called a *ba*, was made of both physical and spiritual elements. The *ba* went on a long and perilous journey, guided by instructions in the *Book of the Dead*. Although the modern name for these scriptures is the Book of the Dead, the translated Egyptian title is more poetic. It was *Coming Forth into the Day*.[3]

A successful *ba* eventually reached a judgment hall where the person's *ab* or heart, representing the character of the person, was put on scales and weighed against a feather. A good person's *ba* went on to paradise, while an evil person's *ba* was destroyed by a monster. A person's deeds were supposed to have an influence on the outcome of the judgment. In the Book of the Dead there is a declaration of the good character of a person, called the Negative Confession. The person declares that he or she has not performed a list of forty-two evil deeds. The following is a version of the Negative Confession from about 1600 B.C.E., published in Homer W. Smith's book *Man and His Gods*.[4] This list gives a clear picture of Egyptian ethics around 3,600 years ago.

1. I have not done iniquity.
2. I have not committed robbery with violence.
3. I have done violence to no man.
4. I have not committed theft.
5. I have not slain man or woman.
6. I have not made light the bushel.
7. I have not acted deceitfully.
8. I have not purloined the things which belonged to the god.
9. I have not uttered falsehood.
10. I have not carried away food.
11. I have not uttered evil words.
12. I have attacked no man.
13. I have not killed the beasts which are the property of the gods.
14. I have not eaten my heart [i.e., done anything to my regret].
15. I have not laid waste plowed land.
16. I have never pried into matters.
17. I have not set my mouth in motion against any man.
18. I have not given way to anger concerning myself without cause.
19. I have not defiled the wife of a man.
20. I have not committed transgression against any party.
21. I have not struck fear into any man.
22. I have not violated sacred times and seasons.
23. I have not been a man of anger.
24. I have not made myself deaf to words of right and truth.
25. I have not stirred up strife.
26. I have made no man to weep.
27. I have not committed acts of impurity or sodomy.
28. I have not eaten my heart [a repeat of number 14].
29. I have abused no man.
30. I have not acted with violence.
31. I have not judged hastily.
32. I have not taken vengeance upon the god.
33. I have not multiplied my speech overmuch.
34. I have not acted with deceit, or worked wickedness.
35. I have not cursed the king.
36. I have not fouled water.

37. I have not made haughty my voice.
38. I have not cursed the god.
39. I have not behaved with insolence.
40. I have not sought for distinctions.
41. I have not increased my wealth except with such things as are my own possessions.
42. I have not thought scorn of the god who is in my city.

The number of statements corresponded to the number of gods in the judgment hall. The need for such a large number led to some redundancy and repetition. The list supports gentleness, reverence, discretion, and truthfulness. Much of the list is similar to the *Ten Commandments* and to morality as we know it. Some of the items would be useful *additions* to our own moral codes. Most of the Ten Commandments would have seemed quite reasonable to Egyptians living a few centuries before the time of Moses.

A well-known Egyptian quotation from about 2200 B.C.E. advises that the temple sacrifices do not count for as much as proper living. "More acceptable is the virtue of the upright man than the ox of him that doeth iniquity."[5] Inscriptions on Egyptian tombs of the Pyramid Age show that people of that time were striving for righteousness. One tomb from about 2600 B.C.E. describes the motivation: "I desired that it might be well with me in the great god's presence." Another tomb said that the owner had "fed the hungry, clothed the naked, fed the wolves of the mountains and the fowl of the sky; that he was beloved of his father, praised by his mother, excellent in character to his brother and amiable to his sister."[6]

The desire to be accepted into paradise at the time of judgment, rather than to be killed and devoured by a monster, was a major motivation for righteous living. Evidence for this is found within the *Memphite Drama*, written near the time of *the origin of writing* in about 3000 B.C.E. It states: *"As for him who does what is loved and him who does what is hated, life is given to the peaceful and death is given to the one bearing guilt."*[7]

The five-thousand-year-old statement given above combines the Golden Rule and the promise of eternal salvation, ideas that were presented three thousand years later in the New Testament. The Egyptian ideas of eternal life, resurrection, judgment, and morality resemble ideas that are also expressed within the Bible. The Bible was not making startling new revelations in presenting these ideas.

EGYPT AND MOSES

Akhenaten, or *Amenhotep IV,* was the pharaoh of Egypt from 1350 to 1334 B.C.E. Unlike his predecessors who believed in many gods, Akhenaten supported a religion having only one god, named *Aten,* or *Aton.* He was referred to as "the only God, beside whom there is no other."[8] He was believed to be the sole creator of the universe. Although Aten was associated with the disk of the sun as well as his temple at *Akhetaten* and the pharaoh himself, Aten was also supposed to be a universal spirit who is present everywhere.

Akhenaten tried to make the cult of Aten the exclusive religion of Egypt. He ordered the names of other gods to be removed from temples. Akhenaten's reign, however, was short and his successor returned Egypt to its ancient religion, with its many gods. The cult of the one god, however, was probably not immediately extinguished after Akhenaten's reign.[9] The cult's beliefs may have influenced Moses, who probably lived in Egypt within the century following the reign of Akhenaten. This was of crucial importance since Moses was fundamental in establishing the nation of Israel and its religion.

William Foxwell Albright (1891–1971), a prominent American archaeologist and educator who specialized in studies of biblical sites in the Middle East, stated:

> In the light of the now available data it is perfectly clear the period between 1350 and 1250 B.C.E. was ideally suited to give birth to monotheism, since it covers precisely the century from the attempted suppression of the solar monotheism of Akhenaten to the middle of the reign of Ramesses II, the great amalgamator of Egyptian and Semitic culture.[10]

Albright placed the date at which Moses led the Hebrew Exodus from Egypt at about 1280 B.C.E., fairly early in the reign of the pharaoh Ramses II. According to Albright, the name *Moses* is of Egyptian origin. Moses is traditionally believed to have held an important position in Egypt. He would presumably have been quite familiar with Egyptian culture, including its religions. Albright suggests that Egyptian beliefs about Aten most likely influenced Moses' beliefs about the nature of *Yahweh. Yahweh* was the name used for God among the Jews during and after the time of Moses.

Points of similarity between the Hebrew Yahweh and the Egyptian Aten are: the concept of one god, recognition of this god's *universal or international dominion,* and taking this *monotheism* as an essential part of the religious system. The case in favor of Egyptian influence on the Hebrew ideas about God is weakened by the fact that some of the beliefs were probably accepted by the Hebrews *before and after* the life of Moses. We should be cautious, however, in accepting scriptural descriptions of Hebrew beliefs *before* the time of Moses. If there was a major shift toward monotheism in Hebrew beliefs at the time of Moses, any *polytheistic* beliefs of the earlier patriarchs might have been minimized in what was written during or after Moses' life.

Although Moses shared some beliefs with the *cult of Aten,* it seems that he was *not* impressed by the large number of the Egyptian gods in the *traditional polytheistic Egyptian religion,* nor by their bizarre images in temples and graves. The practices of worshiping many gods and their images were expressly forbidden in the Ten Commandments that Moses presented to the Hebrews. Most of the remaining Ten Commandments are covered, at least approximately, by items 2, 4, 5, 9, 17, 19, 22, 38, and 42 of the Egyptian Negative Confession given in the previous section.

The similarities between some Egyptian beliefs and the beliefs held by Moses are not sufficient to make us sure that Egyptian beliefs influenced Jewish beliefs. They do show, however, that many of Moses' beliefs resembled those of a neighboring people, suggesting that Moses developed his important teachings by an entirely natural process.

ZOROASTER: THE IRANIAN CONNECTION

Zoroastrianism was a major religion of the ancient Persians. It is especially interesting because so many of its beliefs about supernatural topics resemble those of Judaism and Christianity. *Zoroaster*, or *Zarathustra*, was the religion's founder. A profoundly original religious thinker, he rejected the worship of most gods of the previous religion except for one, the creator of all and the origin of all goodness. This god is named *Ormazd* or *Ahura Mazda* ("the wise lord"). Zoroaster also believed in an evil spirit who opposed Ormazd, producing a universal conflict between good and evil.

It is usually held that Zoroaster lived in Persia during the seventh and sixth centuries B.C.E., but we are not sure of this location or time interval.[11] According to the scriptures, an archangel appeared to Zoroaster in a vision and led him to a conference with Ormazd. After six more visions, Zoroaster began trying to convert people to his religion. In the end, his beliefs became the official religion of Persia. If the Greeks had not defeated the Persians at Marathon, Salamis, and Plataea, this religion might have spread into Europe. This faith still has some followers in Iran and in India. In India Zoroastrians live in or near Bombay and are known as *Parsis*.

Ormazd is supposed to be an all-knowing god who is present everywhere and forever. Ormazd has a complex relationship with his *holy spirit*, *Spenta Mainyu*, who is one with him but is also a distinct entity. In creating the world, Ormazd acted through Spenta Mainyu. It is also through the holy spirit that Ormazd enters the heart of a righteous person and strengthens the person in his or her virtue.[12]

In the struggle between good and evil, the good side is led by Ormazd, who is assisted by bands of good spirits, analogous to angels. The evil side is led by *Ahriman*, aided by bands of demons. The battle is also carried on by the religion's members, who try to do good and fight evil. Their free choices determine whether they will go to heaven or hell. Their evil deeds can be forgiven by confession or the transfer of merit from saints.

The role of *Ahriman* in Zoroastrianism is similar to that of the later Satan of Christianity. Ahriman seduced the first pair of humans to sin. He brought death and numerous demons into the world. The Zoroastrian concept of Ahriman may have influenced Jewish, and consequently Christian, beliefs about Satan. In Jewish scriptures originating *before* the Babylonian Exile, Satan was an agent of God. For example, Satan was empowered by God to torment Job. After the Exile, when the Jews may have been exposed to Persian ideas about Ahriman, Satan became a powerful and independent force opposing God.[13]

Zoroastrianism maintains that humans have a body and a soul. *The soul includes faculties of reason, consciousness, conscience, free will, and a guardian spirit.* The guardian spirit helps good to prevail. Free will enables people to choose good or evil actions and makes them morally responsible for their actions. Since people make their own choices, Ormazd is justified in rewarding or punishing them by sending them to heaven or hell.

Zoroastrianism is waiting for a *savior* or *messiah*. The savior is to be born of a virgin mother who conceives him in a miraculous way. This savior will bring immortality to the righteous and will participate in the destruction of the powers of evil. At the end of the world, Ahriman will direct his hosts in a final battle and will be defeated. The savior will come, and there will be a *resurrection of the dead* and a *final judgment*. In the resurrection, the bones of the dead will be raised up. The final judgment will be a fiery ordeal in which all humanity will be immersed in a river of molten metal in which the just will be saved by divine intervention. Good will be in control forever after, in the "Good Kingdom." The souls of the damned will eventually be purged of the effects of their sins and will participate in the renewed world.

In early Judaism, the Messiah was to be primarily a political leader of the nation of Israel rather than a spiritual leader of universal significance. Even in the earliest teachings of Zoroastrianism, however, the messiah was primarily a *spiritual* leader who would grant immortality to the righteous. Later, Judaism developed the idea of a savior that was closer to that of the Persians, suggesting that the Zoroastrian belief in a spiritual savior influenced the later

Jewish belief.[14] Indirectly, through Judaism, the Persian belief in a savior may have influenced the Christian belief, which greatly resembles the Persian belief.

Zoroastrians believe in frequent prayer. The faultless recitation of standard prayers, sometimes with numerous repetitions, is thought to have a nearly magical power. Prayers are a daily duty and a means of ritual purification. Zoroastrian *purification laws* and rites are even more rigorous than those of the Jews. Many of these practices are very similar in the Zoroastrian and Jewish religions. It has been suggested that Persian influence is responsible for the great importance of purification rites in Judaism.[15]

Like Moses, Zoroaster received his law from his god on a mountain. Zoroastrianism forbids images of its god, and it steadfastly rejects the beliefs of other religions. It has highly developed personal morals and ethics. Morality consists of conforming to the holy will of Ormazd. Before the time of frequent contact of the Jews with the Persians, Judaism emphasized sins of the entire society. *After the lengthy period of Persian domination over the Jews, Judaism placed more emphasis upon personal morality.*[16]

We have seen numerous similarities between Zoroastrianism and Judaism. Starting at the time of the Babylonian Exile, the Jews were in contact with the Persians for over two hundred years. During most of that time the Jews were under Persian domination. Many religious changes occurred in Judaism during this time interval. The Persian rule was relatively benevolent and Jewish thinking may have been open to Persian influence at that time.

It is not possible to prove that the Persian religion influenced the Jewish religion. "Nevertheless, it is hardly conceivable that some of the characteristic ideas and practices in Judaism, Christianity, and Islam came into being without Zoroastrian influence."[17] The Persian origin of numerous Jewish beliefs was also acknowledged by Dr. Kaufman Kohler, a former president of Hebrew Union College, in his book *Jewish Theology: Systematically and Historically Considered.*[18]

Mary Boyce, in her book *Zoroastrianism: Their Religious Beliefs and Practices*, also argued that there had been Persian influence on

Jewish and Christian beliefs. She points out similarities of beliefs, such as that

> there is a supreme God who is the Creator; that an evil power exists which is opposed to him, and not under his control; that he has emanated many lesser divinities to help combat this power; that he has created this world for a purpose, and that in its present state it will have an end; that this end will be heralded by the coming of a cosmic Saviour, who will help to bring it about; that meantime heaven and hell exist, with an individual judgment to decide the fate of each soul at death; that at the end of time there will be a resurrection of the dead and a Last Judgment, with annihilation of the wicked; and that thereafter the kingdom of God will come upon earth, and the righteous will enter into it as into a garden (a Persian word for which is "paradise"), and be happy there in the presence of God for ever, immortal themselves in body as well as soul.[19]

Boyce asserts that "these doctrines all came to be adopted by various Jewish schools in the post-Exilic period."

Although I have pointed out only similarities between the Persian and the Jewish or Christian beliefs, *most of the beliefs of these religions differ from each other.* The similarities might have developed without the actual borrowing of Zoroastrian beliefs by Christians and Jews. There is a problem in showing that Zoroastrian beliefs influenced later Christian and Jewish beliefs since extremely old Zoroastrian texts have not been found and we cannot be sure that all Zoroastrian beliefs are as old as the religion itself. Still, it seems likely that the theme of the final judgment and many other supernatural beliefs held by Christians are somehow related to or descended from earlier Zoroastrian beliefs.

THE MYSTERY RELIGIONS

In the fourth century B.C.E., *Alexander the Great* conquered the Persians and founded an empire that extended from north of Greece in

Macedonia to the Indus River valley in India. He spread Greek culture, Greek cities, and Greek schools throughout this empire. Greek became the language of commerce. People of many nationalities moved from their native regions to settle in other areas within the empire. Even after Alexander died, much of his empire continued to be ruled by people who maintained this Greek lifestyle. This period, before the conquest by the Romans, is called *the Hellenistic Age.*

During the Hellenistic Age, many people lived away from their home cities and the temples of their local gods. To some extent, such people thought of themselves as belonging to humanity in general rather than to a particular city or nation. (Note that the word "cosmopolitan" is of Greek origin.) At the same time, these people had to depend upon *themselves*, rather than their extended family or tribe, and so they became more concerned with their existence as *individuals*.

These people were also exposed to the questioning, skepticism, and theories of the Greek philosophers. Some philosophers held beliefs that seemed threatening to the growing feelings of individuality. One system of beliefs was the *materialism* founded by Democritus and others, which seemed to reduce human existence to a predetermined and mechanical process. Belief in one's *unavoidable fate* became common, as did using *astrology* in an attempt to learn that fate. Many people were slaves, and for them it seemed that there was little joy or justice in the present world.

Perhaps because of the reasons given above, universal religions developed. They offered brotherhood or sisterhood to all who were willing to become members, ignoring differences of social class or national background. Because these religions could accept anyone as a member, the religions could grow almost without limits. Many of these religions offered the individual hope of escaping the iron chains of fate by magic, by willpower, or by entering into special association, or communion, with certain gods. By promising *eternal life*, some religions offered the hope that a happier life could follow life on earth.

During the last few centuries B.C.E., the Greeks were quite religious and worshiped the many gods who are vaguely familiar to us from our acquaintance with Greek mythology. The inhabitants of certain regions had elaborate beliefs and practices honoring a particular god of local importance. These beliefs and practices were nearly separate religions. Often the rites and beliefs of the cults were kept secret. Such cults are called *mystery cults* or *mysteries*.

These new religions spread around the eastern Mediterranean during the Hellenistic Age. Later they spread throughout the Roman Empire. Although some started as cults of local gods, such as the *Eleusinian cults* and the cult of *Dionysus*, others combined characteristics of the Greek cults with spiritual ideas from the Near East. One example of this was the *Isis cult*, which spread from Egypt to other countries. Another very interesting example is the cult of *Mithra*, which probably originated in Asia Minor or Persia and spread into the Roman Empire. This cult will be discussed in the next section.

In the city of *Eleusis*, not far from Athens, a cult developed honoring the goddess *Demeter*. The name Demeter meant "earth-mother," and she was the goddess of fertility and of the harvests. In the myth of Demeter, her daughter, *Persephone*, was abducted by *Hades*, the god of the underworld, and held in the underworld during the barren months of late autumn and winter of each year. In spring Persephone returned, bringing fertility to the fields.

New members of the *Eleusinian cult of Demeter and Persephone* were initiated in rites that included a sacred purifying bath in the sea. In a secret ritual in the temple, members of the cult reenacted Demeter's search for her abducted daughter. Persephone's trip to the underworld and her return symbolized the annual burial of the grain in the soil and its growing to form new grain for the next year's bread. In deeper symbolism, it represented people's hope for immortality after death. The pre-Christian rites of the Eleusinian cult also included a ritual meal of bread and wine.

Another Greek cult centered upon *Dionysus*, the god of wine, who was also a fertility god and a god of trees. He was sometimes represented by a phallus, a bull, or an upright post covered by a bearded mask and a cloak. In a primitive form of *communion*, cult members tore apart an animal that represented Dionysus and ate the raw flesh in the belief that this was equivalent to *eating the flesh and drinking the blood of their god*.[20] By doing this, the believers hoped to be possessed by Dionysus and to become one with him.

Certain beliefs and practices of the Christian religion had previously been maintained by many pagan cults. Most of the cults had initiation rites. As in the cult of Demeter, some of these rites were similar to the Christian *baptism*. Frequently their rites included a ritual meal, commonly including bread and water or wine, that was thought of as *a form of communion with their god*.

Members of some cults, including those of Dionysus, Hercules, Adonis, Attis, and Osiris, worshiped their gods as great benefactors of humanity. Some of the cults celebrated their hero's partial victory over death.[21]

Some pre-Christian gods were also famed for lesser miracles, such as *healing the sick* by supernatural power. There were miracle stories praising Dionysus, Apollo, Demeter, and Hercules. The most famous healer god was *Asclepius*. As an abandoned infant, he was discovered by a shepherd who observed a brilliant light shining from the baby's face. Asclepius quickly became famous for his healing powers. He was even able to *restore the dead to life*. Hundreds of miracle stories of the pagan Greek and Roman traditions have been collected, and they are similar to those later attributed to Jesus.[22]

THE CULT OF MITHRA AND CHRISTIANITY

Mithra or Mithras was the ancient Persian god of light and wisdom. After Zoroaster established something close to monotheism in the Persian religion, Mithra was demoted and became the chief of the good spirits. Later, however, *Mithraism* developed. It was a cult in which Mithra was revered as *the sun god*, or, alternatively, as the conqueror and ally of the sun god. The sun god was previously worshiped in Egypt, Babylonia, and Greece, and this must have helped the Mithra sect spread rapidly in those parts of the world.

According to *Plutarch*, a Greek biographer of the first and second centuries C.E., Mithraism was brought to Rome about 67

B.C.E. by captured pirates from the province of *Cilicia*, on the southeast coast of Asia Minor. The capital of Cilicia was Tarsus, Saint Paul's native city. Mithraism spread rapidly throughout the Roman Empire. It was especially popular among soldiers, slaves, and merchants. Many underground temples of Mithra have been found, even as far west as Germany and England. If Christianity had not developed, Mithraism might have become the dominant religion of the Roman Empire. It was admired by some of the emperors and was briefly the state religion under *Aurelian* in about 273 C.E.

Although it is nearly impossible to learn the precise beliefs held by extinct, secret cults, we do have some idea of the beliefs of Mithraism. Mithra was believed to have been miraculously born out of a rock. This was observed by *shepherds, who brought gifts and adored him*. Mithra tested his strength in various struggles. He overcame, and then allied himself with, the sun god. He killed a wild bull that had been the first animal created by Ormazd. From the body of the bull came all the useful plants and animals of the world. Thus *the death of the sacrificial bull enriched the whole world*.

At the time of a great drought produced by Ahriman (the Satan of Persian religion), Mithra fired his arrows at a rock, liberating a lifesaving stream of water. Then Ahriman produced a great flood, but Mithra advised a man (the equivalent of Noah) to build an ark, and so on. At the end of Mithra's heroic exploits on earth, he held a *last supper with twelve companions*, including the sun god, celebrating their common ventures. Then the gods *ascended to heaven*, with Mithra traveling by chariot, a classy way to travel at that time. From heaven he continued to protect his faithful followers on earth.[23]

Although the life of Jesus was very different from the *story* of Mithra, many of the *rites* taken up by the Christian Church were remarkably similar to those of the Mithra cult. In Mithraism, baptism was called the *sacramentum*. For them, baptism was a form of initiation and purification. It was believed to remove a person's sins, so it could be repeated as needed, as in the Roman Catholic sacrament of confession. At least three distinct rites of baptism were performed.[24] One was baptism by total immersion in water, as was done by the cult of Isis and some Christians.

Another baptismal rite was the sprinkling of *holy water* and making something that was described as similar to the mark of the cross on the person's head. This was probably the cross-shaped sign of the *Sword of Mithra*. It was easy, of course, for early Christians to mistake the mark of a sword for the mark of a cross.

To us, the third method of baptism seems quite bizarre. A nude person would go down into a pit underneath a large grill. A bull would be brought onto the grill and its throat would be cut. The person underneath would be showered in the blood of the bull. In Mithraism, the bull was the symbol of life, and his blood was the most effective method of purification. As the redeeming blood of the dying bull showered the initiate, he was *"born again for eternity."*

There was a Mithraic form of *confirmation*, by which the believer received power to combat the spirits of evil. Analogous to the Christian mark on the forehead in anointing with oil, this ceremony featured a mark on the forehead *by a red-hot iron.*[25] This was not a religion for the fainthearted! Only men could belong to this cult. *Mithra was not a jealous god, however, and women could belong to one of the other cults.*

The Mithraic rite of the *Last Supper* had some features similar to the later Christian Lord's Supper. It was a kind of love feast commemorating the banquet of Mithra that preceded his ascension into heaven. *A celebrant consecrated bread and water or wine.* The bread was in the form of small loaves, each marked with a cross. (Was this the Sword of Mithra again?) From this rite the participant expected to receive health, prosperity, power to combat evil spirits, and immortality in heaven.[26] From their point of view, it was definitely worth doing!

There are some other connections between Mithraism and current practices. Members referred to each other as "brothers" and were initiated by "fathers." The chief of all fathers, the *Pater Patrum*, held his post in Rome. Each day of the week was dedicated to a planet, the sun, or the moon. The most revered astronomical object, the sun, was honored on what they (and Christians) call *the Lord's Day*. We still call the day Sunday. In the case of Mithraism, the lord and the sun had a very close connection, so that the Lord's Day was necessarily sun-day. Early Christianity chose to follow the Roman practice of worshiping on Sunday, rather than Saturday, the Jewish Sabbath.

The birthday of Mithra, called *Natalis Invicti*, was celebrated on December 25, as is the birthday of Jesus. In the case of Mithra, this date made sense in terms of astronomy and Mithra's association with the sun. December 25 is just after the winter solstice and is about when the sun could first be detected moving north from its most southerly position. Thus, this date marked the beginning of the apparent movement of the sun toward northern skies, and was the obvious choice for the symbolic birthday of the sun.

In this and the preceding sections on mystery cults and Zoroastrianism we have seen many parallels with Christian supernatural beliefs and religious rites. Many of these beliefs and rites were not part of Judaism, to which Christianity traces its origin. If we can stretch our imaginations beyond the limits imposed by commonly held beliefs, we can picture a process by which Christianity assimilated and incorporated much of the pagan belief system within itself.

Very early Christians believed they were able to receive divine inspiration and speak in the name of Jesus. With this belief, they could elevate the ideas from their dreams or their personal thoughts to the level of revelations from God. Many early Christians were more familiar with pagan beliefs than with Jewish beliefs. These Christians contributed these pagan ideas to the body of Christian beliefs.

Similarly, the early Christians created their own rites, such as baptism, the Lord's Supper, and so on. Since these rites formed gradually, and more gentiles than Jews became Christian, it was natural that pagan practices were incorporated into the rites since *this was religion as the gentiles had previously known it*. Even Paul, who described himself as a Jew born in Tarsus, lived in an environment in which he

> had more direct access to the concepts and practices of Mithraism than he ever had to the actual words and teachings of Jesus Christ, for in his time none of the four Gospels had been written, and he had to rely on information given to him by Christian converts or by an occasional eye-witness, such as Peter.[27]

Since Judaism is an immediate ancestor of Christianity, I have not tried to describe the many connections between Christianity and Judaism. Many Christian beliefs and practices, however, seem to have had pagan origins, or were drawn from the same pool that was the source of pagan mythology. Long ago, Christians forgot this pool existed.

NOTES

1. Homer W. Smith, *Man and His Gods* (New York: Grosset & Dunlap, 1952), p. 117.

2. William Foxwell Albright, *From the Stone Age to Christianity: Monotheism and the Historical Process*, 2d ed. (Baltimore: Johns Hopkins Press 1957), p. 238.

3. Smith, *Man and His Gods*, p. 40.

4. Ibid., pp. 43, 44.

5. John Bartlett, *Bartlett's Familiar Quotations*, 14th ed. (Boston: Little, Brown and Company, 1968), p. 3.

6. Smith, *Man and His Gods*, p. 49.

7. Ibid., p. 46.

8. Albright, *From the Stone Age to Christianity*, p. 221.

9. Ibid., p. 223.

10. Ibid., p. 12.

11. Mary Boyce and Frantz Grenet, *A History of Zoroastrianism*, 3 vols. (Leiden: E. J. Brill, 1991), 3:370.

12. Ibid., p. 425.

13. Jim Hicks, *The Persians* (New York: Time/Life Books, 1975), p. 105.

14. George William Carter, *Zoroastrianism and Judaism* (Boston: Gorham Press, 1918), p. 80.

15. Ibid., p. 91.

16. Ibid., p. 94.

17. S. A. Nigosian, *The Zoroastrian Faith* (Montreal: McGill-Queen's University Press, 1993), p. 97.

18. Kaufman Kohler, *Jewish Theology: Systematically and Historically Considered* (New York: Macmillan, Company, 1923), pp. 140, 184, 185, 288, 301.

19. Mary Boyce, *Zoroastrians: Their Religious Beliefs and Practices*, rev. ed. (London: Routledge and Kegan Paul, 1986), p. 77.

20. Smith, *Man and His Gods*, p. 127.

21. Shirley Jackson Case, *The Origins of Christian Supernaturalism* (Chicago: University of Chicago Press, 1946), pp. 79–84.

22. Burton L. Mack, *Who Wrote the New Testament? The Making of the Christian Myth* (New York: HarperCollins, 1995), p. 65.

23. Franz Cumont, *The Mysteries of Mithra*, trans. from the 2d French ed. (New York: Dover Publications, 1956), pp. 130–40.

24. Esmé Wynne-Tyson, *Mithras* (London: Rider & Company, 1958), p. 43.

25. Cumont, *The Mysteries of Mithra*, p. 157.

26. Ibid., p. 160.

27. Wynne-Tyson, *Mithras*, p. 70.

4

NEW IDEAS ABOUT THE
ORIGIN OF THE NEW TESTAMENT

MANY PEOPLE IMAGINE THAT THE disciples of Jesus witnessed astonishing miracles. Recently, many biblical scholars are painting a very different picture. They say that the disciples did not see the supernatural events reported by the Bible, and that the reports of miracles and other wonders are stories that developed *after* the life of Jesus.

Trusting the Bible leads many to accept ancient beliefs about the soul. If there are good reasons to believe that the Bible's supernatural events didn't really happen, there may also be good reasons to doubt the doctrine of the soul. In this chapter, I argue that the supernatural features of the Bible stories are fictitious.

THE BOOK OF Q

Since there is very little ordinary historical information about Jesus, the Gospels are our main source of knowledge about him and his teachings. Because of this, they are the most important part of the New Testament. Mostly, I will discuss the *synoptic Gospels*, those of Matthew, Mark, and Luke. These names will refer to the Gospels or to the authors of those Gospels. You will know which is meant from the context.

Of the four Gospels in the Bible, Matthew is presented first, and many people believe that Matthew was the first one written. However, scholars now believe that Mark was first. It was written about 70–80 C.E.

In itself, the order of the Gospels is not so exciting, but the situation becomes more interesting if we ask whether Matthew and Luke used Mark's Gospel in writing their Gospels. It appears that they did, since they have much of the same material as Mark, and much of that material is presented in the same order that Mark presented it. *However, a great deal of material is present in both Matthew and Luke that is absent in Mark.* That material, called "Q," consists primarily of *the sayings of Jesus.*

The Q material, sometimes known as "the Sayings Gospel Q," got its name from the initial letter of the German word *Quelle,* meaning "source." It was proposed that Q and Mark were the two source documents for Matthew and Luke. Various scholars, including the German Lutheran scholar and theologian Rudolf Bultmann (1884–1976), found strong support for this idea. Since the 1920s it has been widely accepted that Matthew and Luke had at least two sources: Mark and Q.

Evidently, Q was available to Matthew and Luke, but no ancient Q text now exists. Part of Q can be reconstructed from the synoptic Gospels. The Q material is identified by its presence in both Matthew and Luke, but not in Mark.

There is another clue in this puzzle. Q is broken into many segments in Matthew and Luke, and these segments are interleaved with Mark's material. Remarkably, though, the Q segments are placed at different locations in Matthew and Luke. This suggests that Matthew is not a copy of Luke or vice versa.

We can understand the different locations of the material in Matthew and Luke. Mark presented his material as if it were in chronological order, so Matthew and Luke usually followed Mark's order for his material. However, Q consists mainly of Jesus' sayings, and *these could be rearranged without affecting their meaning.* Since there was no unique way of weaving the Q material into Mark's material, it is no surprise that independent writers did this in different ways.

Karl Ludwig Schmidt, a prominent biblical scholar, published an important analysis of Mark in 1919. Writing about four decades after the death of Jesus, Mark had no previous biography of Jesus to assist him. Schmidt showed that Mark composed his Gospel by *writing his own connecting links between various smaller stories about Jesus.* This result nearly destroyed the search for the "historical Jesus" in the Gospels.[1] Evidently, details about the life of Jesus, such as his movements in Galilee and Judea, were mostly fiction.

In 1945 there was an important discovery of ancient manuscripts at Nag Hammadi, near the Nile in Upper Egypt. The location had been a library of *Gnostic Christian* documents. The *Gnostics* were a group that held different religious and philosophical beliefs than the mainstream Christians. They believed that they possessed special, divinely revealed knowledge, and that the material world is intrinsically evil. At the end, one of the manuscripts was signed, "The Gospel according to Thomas." The contents are a previously unknown collection of Jesus' sayings. About one-third of the sayings correspond to similar sayings in Q. Other parts of the material are not present in the usual four Gospels.

The *Gospel of Thomas* includes discussions between Jesus and his followers. It promotes Gnostic views by presenting new sayings of Jesus supporting that belief system.[2] This shows that a Christian group with divergent beliefs produced a Gospel that supported the group's own theology. Those who view the scriptures as results of purely natural development processes, of course, are not surprised by this. Even among believers, it is often emphasized that the Gospel writers were trying to support their particular theological views and were not trying to write history.[3]

THE RISING IMAGE OF JESUS BURIED IN Q

Recent work on Gospel origins, based mostly on analysis of the Gospels themselves, describes how Q gradually came to present

Jesus as a supernatural being. Obviously, these conclusions are closely linked with the roots of the Christian faith. Burton Mack, a professor of New Testament at the Claremont School of Theology, presents many of the conclusions in two informative and stimulating books. They are *The Lost Gospel: The Book of Q and Christian Origins* and *Who Wrote the New Testament? The Making of the Christian Myth.*[4]

Another major source of new results about the Bible is *Honest to Jesus: Jesus for a New Millennium.*[5] It is a very readable book by Robert W. Funk, who summarizes the major conclusions of the *Jesus Seminar*, which he founded. The Jesus Seminar is a large group of biblical scholars devoted to uncovering the real *Jesus of history*, who tends to be hidden by the traditional *Christ of theology*.

The book of Q, reconstructed from material that is present in both Matthew and Luke, but not Mark, fills twenty-two pages in Mack's *The Lost Gospel: The Book of Q and Christian Origins.*[6] Q consists of sixty-two sayings, which Mack labels QS 1–QS 62. (In some following references to material from Q, the sayings are represented by a QS number from Mack, as well as their location in Luke. Unlike the Gospels, Q does not present the story of Jesus' life. Most of Q is made up of the *sayings of Jesus*. Analysis of these sayings showed they are similar in type and style to compilations of wise sayings, or *wisdom sayings*, in other early Christian writings and in earlier Hebrew and Egyptian literature.

The sayings depict Jesus as a loving and consoling teacher, as well as a countercultural leader urging his followers to give up their traditional way of life to imitate his homeless lifestyle. Seven of the sayings mention *the kingdom of God*, evidently a theme of major importance, although its precise meaning is uncertain. (We saw the phrase "kingdom of God" in the previous chapter. Mary Boyce mentioned it among other concepts that may have originated with the Zoroastrians.)

There is more to Q than the wise sayings of Jesus. Many segments show Jesus as a prophet, predicting that those who do not follow his instructions will be sorry on judgment day. He predicts that great disasters will precede the coming of the "son of man" and the judgment. This prediction of a disastrous end of the world, followed by judgment, is called the *apocalyptic theme*, and coincides with beliefs that most likely originated with the Zoroastrians and later were held by many Jews. In the Old Testament, such beliefs were present in the Book of Daniel.

Scholars were disturbed that Q contains writings of both the wise-sayings type and the apocalyptic type. In religious literature these two kinds of writing are usually not done by a single author, since they are likely to be produced by different types of personalities. The situation became more complicated about 1988, when some scholars concluded that *there are three levels of material in Q*, denoted by Q^1, Q^2, and Q^3. These levels are discussed in Mack's *The Lost Gospel*.

The different levels of Q were written at different times. This picture was first proposed by a biblical scholar named John Kloppenborg in 1987.[7] Level Q^1 is the oldest part and consists of the wise sayings of Jesus. Q^2 was written next. It includes the prophetic and the apocalyptic material. It is believed that the Q^1 material was written before there was knowledge of (or interest in) the Q^2 material. However, *the Q^2 material depended upon and was organized around the earlier Q^1 material*.[8] Finally, a small amount of material was added after Q^2, called Q^3.

Not only is Q^1 thought to be the first part of Q to be written, but it is thought to contain the *earliest known views* of the group, called the *Jesus movement*, that gave rise to Q. Many biblical scholars believe that Q^1 is the most likely source of information about the "real Jesus." The later writings, including the Gospels, are thought to contain many fictitious stories that were composed to fit later beliefs about Jesus. Mack notes that the Jesus movement, "did not know about or imagine any of the dramatic events upon which the narrative Gospels hinge."[9]

How later writers felt free to add their own material to the religious literature will be discussed later. Like Q^1, Paul's writings were very early, but Paul never saw Jesus in person, and he gives little information about the life of Jesus. It is likely that many of the Gospels' details about Jesus' life *were not known by anyone* at the time of Paul's writings.

Q¹ portrays Jesus primarily as a wise teacher. In Q¹, *Jesus does not exhibit, and does not imply that he has, supernatural powers.* He does not perform miracles or predict the future. He does not imply that he will return to earth before the last judgment. There is no statement that he is a redeemer or messiah. He refers to God as a separate entity and not as his own father. No instructions are given to Peter and the other disciples to found a church, and there is no baptizing of converts. These ideas were added later and quite likely were not held by Jesus.

The "kingdom of God" in Q¹ appears to refer to a movement to spread Jesus' views. The kingdom was compared with the growth of a mustard seed in a garden or yeast in bread dough. This suggested that it was capable of enormous growth. Mack holds, contrary to a common opinion, that *the earliest idea of the kingdom was not connected with an apocalyptic view of coming events.* (QS 46, Luke 13:18–21)

Mack believes that the Jesus people suffered major setbacks in their goal of establishing the kingdom of God, leading to changes in their thinking, exhibited in Q². This level of Q criticizes groups that had rejected the movement. Q² condemns Pharisees, those who disown Jesus in public, "this wicked generation," and the towns of Chorazin, Bethsaida, and Capernaum. Jesus promises the town of Capernaum that at judgment, "You will be told to go to hell." John the Baptist sets the harsh tone of Q² at the start. Addressing people who had come, seeking to be baptized in the Jordan River, John shouts, "You offspring of vipers!" (QS 37, 34, 32, 22, 4; Luke 12:9, 11:39–44, 11:29, 10:13, 10:15, 3:7)

Perhaps responding to the need to strengthen their message, the Jesus movement introduced new teachings about Jesus in Q². The changes must have bothered those familiar with the previous Q¹ material. Anticipating this, the writers had Jesus say, "And fortunate is the one who is not disturbed [at hearing these things] about me" (Mack's brackets). (QS 16; Luke 7:23)

The new teachings about Jesus raised his image from a wise teacher to a being with supernatural powers. Jesus became a *miracle worker* in Q² with a story of a long-distance cure of a paralyzed and nearly dead servant of a Roman army officer. He also exorcised a demon, allowing a speechless person to speak. (QS 15, 28; Luke 7:1–10, 11:14)

In speaking about himself to disciples of John the Baptist, Jesus is made to say, "Go and tell John what you hear and see: the blind recover their sight, the lame walk, lepers are cleansed, the deaf hear, the dead are raised, and the poor are given good news." This statement paraphrases predictions in Isaiah that such things would happen at the time of the restoration of the Kingdom of Israel. This may have been used as a list of preferred topics for the miracle stories that appeared later and were incorporated into the Gospels. These stories then became part of the "evidence" that Jesus fulfilled Isaiah's prophecy. (QS 16; Luke 7:19–22)

A major theme of Q^2 is the announcement of impending judgment. This introduces the Zoroastrian and Hebrew idea of a final judgment into Q. It warns of severe consequences for anyone who rejects the movement's message. It appears to be the movement's harsh response to being rejected by many. (QS 32, 41–45, 60, 61; Luke 11:31, 32, 12:39–40, 42–46, 49–59)

The theme of judgment had been completely absent in Q^1 and the Book of Thomas. Thomas does contain much of the material of Q^1. This suggests that the Gnostic Christians who wrote Thomas broke away from the Jesus movement at least before the end of the Q^2 stage.

The judgment theme enhances the status of Jesus in a number of ways. It declares that ignoring Jesus' words is a big mistake. For the first time, as he predicts coming disasters and judgment, *Jesus appears to be a prophet with knowledge revealed by God, rather than just a wise teacher.*

Once, in Q^1, and seven times in Q^2, Jesus refers to himself using the ambiguous term "the son of man." This term can simply refer to a human being. It can also refer to the "son of man" predicted in the Book of Daniel who will have power over all the earth at the final judgment. In Q^1, Jesus seems to use the term merely to refer to himself, a human being. In Q^2, its use in the apocalyptic sense gradually becomes more evident, until Jesus says, "Every one who admits in public that they know me, the son of man will acknowledge before the angels of God." This strongly links Jesus with the heavenly figure who will preside at the last judgment. (QS 8, 18, 19, 32, 37, 41, 60; Luke 6:22, 7:34, 9:58, 11:30, 12:8, 12:10, 12:40, 17:24, 17:26)

Starting in Q², Jesus is presented as being more than a human being. By careful use of the term "the son of man," it was gradually insinuated that Jesus would possess the power to preside over the activities on judgment day. This gradual approach was probably necessary since previous texts had not depicted Jesus as a possessor of supernatural powers.[10]

From about 66 C.E. to 73 C.E. there was a series of riots and battles in Judea between different Jewish factions and between the Romans and the Jews. The Jesus movement made some additions to Q after the end of the Roman-Jewish war. These additions are called Q³. They include, among others, the temptation of Jesus, the revelation to the "babies" about the son of God, a warning about hellfire, and the lament of Jesus over Jerusalem. In the ongoing process of building up the importance of Jesus, *Q³ makes a case that Jesus is the son of God.*

In the temptation story, Jesus is asked to prove that he is the son of God. He does not comply with the request, but a reader is left thinking that Jesus *may actually be* the son of God. As with the "son of man" in Q², the reader is introduced to this new idea by a gentle suggestion that it may be true. (QS 6; Luke 4:1–13) Next, Jesus says:

> I am grateful to you, father, master of heaven and earth, because you have kept these things hidden from the wise and understanding and revealed them to babies. Truly I am grateful, father, for that was your gracious will.
>
> Authority over all the world has been given to me by my father. No one recognizes the son except the father; and no one knows who the father is except the son and the one to whom the son chooses to reveal him. (QS 24; Luke 10:21–22)

So, although wise and intelligent people do not perceive these truths, God reveals them to children. Then Jesus claims that his father has given him authority over the whole world. His father must be God! At this stage Jesus is effectively declaring himself to be the son of God. As understood by Mack and others, however, this is not really Jesus speaking, but later authors who believed it was appropriate and in character for Jesus to say this.

At first sight, the lament over Jerusalem seems to be a prophecy by Jesus that came true about forty years later during the Roman-Jewish war. However, Q³ was probably written *after the war*. The lament should not be regarded as a fulfilled prophecy, since the "predicted" occurrence had already happened. (QS 49; Luke 13:34, 35)

Although the people who wrote Q had an exalted picture of Jesus by the time they finished Q³, Mack writes that

> the people of Q were not Christians: the people of Q did not think of Jesus as a messiah, did not recognize a special group of trained disciples as their leaders, did not imagine that Jesus had marched to Jerusalem in order to cleanse the temple or reform the Jewish religion, did not regard his death as an unusual divine event, and did not follow his teachings in order to be "saved" or transformed people.[11]

GALILEE AND THE GREEK CONNECTION

Galilee is the setting of most of the New Testament. We tend to think of Galilee as part of Israel, so we think that it must have been mostly Jewish. *However, Galilee was not under the control of Jerusalem from the death of Solomon in 922 B.C.E. until it was taken from the Syrians by the Maccabean dynasty in 104 B.C.E.* It should be noted that there was local opposition to the Maccabean conquest of Galilee. Then, about forty years later, Galilee fell under Roman control.

When Jews returned from exile in Babylon in 539 B.C.E., they returned to Judea, not to Galilee. Jews referred to Galilee as "the land of the Gentiles."[12] Geographically, Galilee was closer to Phoenicia and Syria than to Judea, Judaism's center. Its people included Syrians, Phoenicians, Arabs, Greeks, and Jews. Larger numbers of Jews settled there after the Roman-Jewish war, which ended in 73 C.E.

Galilee was heavily influenced by Greek culture for the three hundred years that preceded the life of Jesus. *During Jesus' life, southern Galilee was mostly Greek speaking*. The second language was *Aramaic*, a language closely related to Hebrew. Aramaic was probably Jesus' native tongue. Many Greek cities were in and near Galilee. Sepphoris, located within an hour's walk of Nazareth, was largely Greek in lifestyle. Across the Jordan River to the southeast of Galilee was the Decapolis, for which the Greek name refers to the ten Greek cities that had been founded there.

Since the Greek way of life was so strongly present in Galilee, we should not assume that Jesus was influenced only by the Jewish culture, especially since there is evidence of Greek culture in the sayings of Jesus in Q^1. Remember that, of all the Gospel material, these are the most likely to represent the actual teachings of Jesus.

In order to discuss the Greek influence, let's consider what Jesus taught in Q^1. Jesus told his followers to sell their possessions, give to charity and beggars, lend without expecting anything in return, and not to worry about food or clothing. He told them to leave their homes and family and to work single-mindedly for the kingdom of God. They were told to trust that God would give them what they needed. They were not to carry money, a bag, sandals, or a staff. They were to knock on doors, treat the sick, and accept food that was given to them. They were to love their enemies. They were not to retaliate if mistreated.

The kind of lifestyle which Jesus advocated and practiced was similar to that of a certain group of Greek sages. They left their families, homes, and possessions; begged for food; and served as critics of those with privilege, power, and possessions. They were called the *Cynics*, but they did not conform completely to our modern use of this word. They lived as lone beggars who challenged the world with only a staff, sandals, a pouch, and a characteristic cloak. Note that Jesus recommended that his followers give up even the staff, sandals, and pouch.

The Cynics were highly regarded for their self-assuredness, independence, witty responses, nonretaliation, and nonconformity. One famous Cynic philosopher, Diogenes of Sinope, went too far in one of his critical comments. His face was slapped. Diogenes then asked himself out loud why he had forgotten to wear his helmet that day. Like Jesus, the Cynics were sometimes criticized for keeping bad company. In response, the Cynic Antisthenes replied, "Well, physicians attend their patients without catching the fever." When someone criticized Diogenes for frequenting unclean places he replied that the sun also enters privies without becoming defiled.[13]

Many scholars have maintained that the sayings of Jesus in Q[1] suggest that he resembled a Cynic sage in his style of interacting with people, as well as in his way of life. The multiethnic crowds who gathered around Jesus in Galilee must have appreciated the Cynic's clever interchanges. Some of the *contents* of Jesus' message are also similar to beliefs of the Cynics. Jesus and the Cynics espoused nonretaliation, challenged social conventions, and advocated unconventional ways of living, but did not work to overthrow the existing society.

Thus, Jesus' way of life and some of his beliefs suggest that he was influenced by the Cynics. Jesus could have had access to Cynic philosophy. A city named Gadara was famous as a source of Cynic philosophers and poets. It was located in the Decapolis about a day's walk from Nazareth. Considering how little we really know about the life of Jesus, he may even have been schooled in Cynic philosophy.

It is often assumed that Jesus taught in Aramaic, but this may be mistaken. The oldest manuscripts of the New Testament are in Greek, or, when they are in other languages, they have clearly been translated from previous Greek versions. Robert Funk refers to evidence that the sayings and parables of Jesus were originally composed in Greek.[14] The poetic quality of the parables and sayings in the Greek manuscripts argues that they had not been translated. A translation would have robbed them of much of their resonant style.

Greek influence may be part of the answer to a question sug-gested by the discussion of the layers of Q. The question is: how could the writers who contributed to the later stages of Q attribute their own fictional material to Jesus? One answer is that this was typical for writers educated in Greek schools.[15] This practice makes it difficult for biographers to decide which material attributed to an ancient writer was actually contributed by the writer himself, rather than by his disciples.

Greek schools were the most influential schools in the Roman Empire. They were numerous in the parts of Palestine and Syria where much of the New Testament originated. Many in the Jesus Movement must have had Greek schooling. These people would have been comfortable attributing their own material to the head of their school of thought. For them, this was a *gift* to their move-ment's founder.

We do not need to appeal to *Greek* influence to explain the prac-tice of scholars attributing their work to their predecessors. Maurice Casey, a scholar concerned with Jewish and Christian origins, has noted that this was also a practice among the ancient *Jewish* writers.[16] Casey points out a number of Old Testament examples, including that the total number of psalms attributed to David num-bers about 3,600. Thus the early Christians might have acquired this habit from *both* the Jewish and the Gentile traditions.

We should not expect that the Gospels escaped having fictional material added to the sayings of Jesus. In fact, biblical scholars have evidence that such additions were made, as in Q^2. The writers who made these additions may never have imagined that they were adding to what many now consider to be the true "Word of God."

By far, most of the Bible-like material that was written about Jesus in the first few centuries of the Christian Era *did not* become part of the New Testament. There is, for example, an amusing mir-acle story in the *Infancy Gospel of Thomas* about the child Jesus when he was six years old. Another child had slapped Jesus in the face. Jesus told the child to "finish his course," and the child promptly dropped dead.[17]

THE CHRISTIAN BRANCH
OF THE JESUS MOVEMENT

The Jesus movement that produced Q[1] developed in Galilee about 50 C.E. By about this time another set of beliefs about Jesus had developed to the north, in northern Syria. These beliefs were radically different from those of Q. They produced a drastic change of direction among the followers of Jesus. Let's follow Mack and call the group that developed these beliefs the *Christ cult*. Although the people who wrote Q emphasized the sayings of Jesus and never mentioned his death and resurrection, the Christ cult concentrated on the significance of the death and resurrection of Jesus and paid little attention to the sayings or the actual life of Jesus. The Christ cult emphasized Jesus' role as a redeemer rather than as a spiritual teacher.

The Christ cult first repelled Paul, but eventually converted him. Paul helped develop and spread the cult's beliefs in Syria, Asia Minor, and Greece. *The cult developed in regions that were heavily influenced by Greek mystery cults*. The theology of this group was influenced by the pagan cults, resulting in *a hero and savior cult devoted to the god Jesus Christ*. As in other cults, membership was open to people of all nationalities. For members, participating in the cult was equated with belonging to the kingdom of God.

Like many cult heroes, Jesus was believed to have had a heroic death, followed by a resurrection from the dead and ascension into heaven. His followers developed ritual practices, such as singing hymns, reciting poetry, and repeating standard prayers. As in the cults of Dionysus, Demeter, and Mithra, ritual meals were celebrated in his memory. Following the practice of many mystery cults, members were initiated by a formal rite called *baptism*.

Like the Jesus movement that produced Q, the Christ cult contributed to a greatly increased status of Jesus. He became a savior who had sacrificed himself to pay for the sins of all humanity. This was meaningful to both gentiles and Jews.

For the Greeks and some other gentiles, Jesus' death fit their picture of *a noble death*. The noble death had originally been the death of a soldier for his country. Later it was applied to the death of a philosopher for his beliefs, as in the case of Socrates. Dying for one's country or beliefs was "the ultimate test of virtue, and obedience unto death the ultimate display of one's strength of character."[18] As mentioned in the previous chapter, the idea of *the death of a savior to save all humanity* was part of the mythology of many pagan cults that preceded Christianity.

The Jews believed that the deaths of the seven Maccabee brothers "purified the land," as stated in 4 Maccabees: "having become as it were as a ransom for our nation's sin." Thus redemption by martyrdom was a valued tradition among the Jews, as well. Among Jews there was also the tradition of the persecution of a righteous person, followed by vindication, as in the stories of Joseph, Esther, and Daniel. This fitted in with the suffering and death, followed by *resurrection as vindication by God,* in the Christ cult beliefs about Jesus.

The Jesus people who wrote Q did not imagine Jesus as a savior, redeemer, or as a resurrected person, and they cannot properly be called "Christians." On the other hand, the people of the Christ cult did not usually think of Jesus as a wise teacher. *These very different traditions were brought together by Mark when he wrote his Gospel. His Gospel helped form a central body of Christian beliefs.* In the Greek tradition of linking together all available stories about a person to make a biography, Mark assembled Jesus stories in a way that included traditions of the Jesus people and of the Christ cult. He wrote about the death and resurrection of Jesus as if it were history, padding the bare bones of the Christ cult's beliefs with layers of detailed fiction.

As emphasized by many, the "Mark" credited with writing the first Gospel was *a fictional name* attached to an unsigned document resulting from a process of writing, polishing, and rearranging.[19] A great deal of literary skill and design is present in all of the Gospels. The common belief that the Gospels were written by Galilean fishermen is not very believable unless one also believes that the writing took place with supernatural assistance.

THE GOSPELS AS BIOGRAPHIES
OF A CLASSICAL HERO

For us, the Gospels are not like anything else that we read. They seem to be a unique form of literature. Remarkably, though, parts of the Gospels follow a pattern that was well known in the ancient world. The pattern started as a stereotyped biography of Greek military heros. These biographies described four basic elements of a hero's life. They were the hero's *birth* or genealogy, a remarkable episode about his or her *youth*, the important *achievements* of the hero as an adult, and a recounting of the hero's *death*. A hero might be known primarily because of one category, but if the hero was truly famous, the stories multiplied and grew until there were impressive entries in several categories.

The pattern was modified when applied to important philosophers or religious figures. The achievements were divided into wise sayings and wondrous deeds. According to Aristoxenus, a student of Aristotle, a standard biography included five categories:

   a. A miraculous or unusual birth;
   b. A revealing childhood episode;
   c. A summary of wise teachings;
   d. A description of wondrous deeds;
   e. A martyrdom or noble death.[20]

This pattern was known to ancient scholars and was called an *encomium*. Together, the Gospels cover all these categories in Jesus' life.

For describing a hero's *birth*, there is a further breakdown of standard features that are included in a so-called *infancy narrative*.[21] According to Lane McGauhy, whom Funk cites, five parts were often found in a classical infancy narrative:

   1. A *genealogy* revealing illustrious ancestors;
   2. An unusual, mysterious, or miraculous *conception*;

3. An *annunciation* by an angel or in a dream;
4. A birth accompanied by supernatural *omens*;
5a. *Praise* or a forecast of great things to come; or
5b. *Persecution* by a potential competitor.

As examples of infancy narratives, Funk describes those of Plato and Alexander the Great. Plato's genealogy was traced back to Solon, a famous Athenian statesman, and Poseidon, the Greek god of the sea. Although Plato's father had stopped making love to his wife because she seemed unable to conceive, she became pregnant. The god Apollo appeared to Plato's father in a dream, presumably to inform the father of Plato's imminent birth. Plato was often portrayed as a son of Apollo. There was also a legend that Plato was born on Apollo's birthday, which was regarded as a favorable omen. Thus, stories about Plato included parts 1 through 4.

The life of Alexander the Great was described by Plutarch in the first century C.E. Plutarch wrote that Alexander was descended from Hercules on his father's side, and from Zeus on his mother's side. Two versions of stories involving snakes implied that Alexander's mother conceived him from a union with a higher being. Both of Alexander's parents had remarkable and promising dreams related to the pregnancy. Notable events occurred on the day of Alexander's birth. An important temple burned down, portending later calamaties. Alexander's father, the king of Macedonia, received news that his army had won an important military victory and that his horse had won at the Olympic games. The king's soothsayers interpreted these victories as signs that Alexander would be unconquerable. Alexander's story included all but part 5b of the infancy narrative format.

It is revealing to note elements that are present or absent in the synoptic Gospels. In the earliest Gospel, Mark starts with the adult achievements of Jesus, with no infancy narrative or revealing childhood episode. Mark presents wise sayings, wondrous deeds, and a description of Jesus' death. *The later Gospels of Matthew and Luke fill in the missing elements of the classical enconium and the infancy narrative.*

Matthew's Gospel includes the elements in Mark, but also includes an infancy narrative. A *genealogy* presents Jesus as a descendent of Abraham and David through Joseph, his father. The inheritance is traced through Joseph, even though Joseph was not portrayed as the biological father of Jesus. In Matthew's *annunciation*, an angel appears to Joseph in a dream, and tells him not to worry, that Mary's *conception* was "of the Holy Ghost."

A favorable *omen* appeared in Matthew's account of the birth of Jesus: a star guided wise men to the birthplace. The wise men brought gifts and worshiped Jesus, an implicit form of *praise*. Later, in a dream, an angel warned Joseph that he needed to take Mary and Jesus to Egypt, to avoid *persecution* by the jealous King Herod. Thus, Matthew's infancy narrative includes all the typical parts.

Most people are familiar with Matthew's charming description of the "star of Bethlehem" guiding the Magi (ancient Persian priests) to Jesus' birthplace. The fictional origin of this Bible story is supported by the following passage in a Zoroastrian holy book, the *Avesta*.

> You, my children, shall be the first honored by the manifestation of that divine person who is to appear in the world: a star shall go before you to conduct you to the place of his nativity; and when you shall find him present to him your oblations and sacrifices; for he is indeed your lord and an everlasting king.[22]

If the Zoroastrian religion were recognized as one of Christianity's parent religions, people would surely regard this as a prophecy about Jesus that was fulfilled at his birth!

Luke's Gospel includes a story about the twelve-year-old Jesus revealing astounding knowledge to the learned men at the Temple in Jerusalem. This is the only *revealing childhood episode* in the Gospels. In Luke, infancy narratives of Jesus and John the Baptist are intertwined. John the Baptist's includes a *genealogy*, an *annunciation* by the angel Gabriel, and a *conception* by an old woman. When John was eight days old, a remarkable *omen* occurred. His father regained his long-lost ability to speak. The father immediately praised God and predicted that John would prepare the way for the Lord.

Luke's infancy narrative of Jesus contains many of the now familiar standard parts, but *the actual episodes are all different from Matthew's. Luke's genealogy of Jesus is highly inconsistent with that given by Matthew.* Luke's *annunciation* is to Mary, not Joseph. The *omens* in Luke involve appearances of angels to shepherds, not a star and wise men. There is no category 5b persecution story in Luke, but the infant receives category 5a *praise* from Simeon and Anna at the circumcision in Jerusalem. All of this suggests that the body of lore about Jesus had become so rich that Matthew and Luke came up with entirely different episodes in their processes of including all the classical elements in their biographies of Jesus.

Many more Jesus stories were produced that were not included in the Bible. These other scriptures filled in the lore about Jesus and others close to him. The *Infancy Gospel of Thomas* (not the same as the Gospel of Thomas) describes Jesus working miracles from infancy until his appearance at the Temple in Jerusalem. The *Infancy Gospel of James* describes the earlier years of Mary, Jesus' mother. It claims that Mary was a virgin, in the normal physical sense of the word, both *before and after* she gave birth. This Gospel also presents a persecution story of John the Baptist, describing a miraculous escape of John's mother from Herod's agents. Some other Gospels not found in the Bible are those of Peter, Phillip, and Mary Magdalene.

THE EVOLUTION OF BELIEFS ABOUT JESUS

At first, Christianity did not have a conservative hierarchy or a standard set of beliefs, and ideas such as the meaning of the kingdom of God were quite vague. In this rather fluid situation, various groups of believers developed. Two groups have been discussed previously, the Christ cult and the Jesus movement that produced Q. There were many other groups, including groups with Gnostic tendencies who produced the Gospels of Thomas and John.

By processes similar to biological evolution, the groups developed new ideas that appealed to themselves and to potential converts living nearby. By random creative processes and the effects of differing situations, the beliefs of these groups began to differ.

Those groups that developed unattractive ideas were not able to maintain and increase their membership. More successful groups grew rapidly. They helped form the mainstream Christian beliefs. The most successful groups developed ideas that offered more to new members than other Christian and pagan groups. *As a result of all this, most members eventually belonged to groups with beliefs that were very attractive to potential converts.* This was a form of "fitness," analogous to evolutionary fitness. By such processes, the very early Christian communities adapted to the environment even without strong guidance from a hierarchy of leaders.

In the competition with the pagan religions, the status of Jesus had to exceed that of all of the pagan gods. Beliefs about Jesus evolved until this was accomplished. By gradual advances made in various groups, Jesus' status passed through many stages: wise teacher, prophet, miracle worker, son of man, son of God, cosmic redeemer, lord of all creation, and one of three persons in God. The "good news" kept getting better and better! New supernatural powers were attached to Jesus that were not known to his earliest followers. Most of Jesus' enhanced status developed early enough to be incorporated into the Gospels.

The cosmopolitan nature of the Roman Empire brought the religions of all nations of the Mediterranean world into competition, giving rise to a *spiritual marketplace.* In this marketplace, a religion could succeed by making lavish claims about what the supernatural world could do for a person. Christianity's offer was hard to beat: it offered people *the possibility of eternal happiness.* The Christian religion proclaimed *a God of unlimited power and unmatched goodness,* greatly surpassing the pagan gods, who displayed many limitations of ordinary humans. Faith in the new religion was made easier by miracle stories that described Jesus' supernatural powers.

JESUS' MIRACLES

Three common reasons why people believe in Jesus are:

1. That he rose from the dead;
2. That he perfomed many amazing miracles; and
3. That he fulfilled the Old Testament prophecies.

If we believe the Bible is literally true, then we accept all of these reasons. If, however, we believe the Bible was written by an ordinary and natural process, then it is possible that reasons 1 and 2 are not true: The miracles and resurrection may be fictitious. In addition, the Bible was written by people who were strongly aware of the Jewish scriptures. A good case can be made that the scriptures were *the basis* for much fictional material in the Gospels. If this is so, the Gospel stories naturally fulfill Old Testament prophecies, *since the Jewish scriptures are the sources of the stories.*

Anyone who has read the Gospels is aware of many references to Old Testament prophecies and to numerous statements that things were done "that the scriptures might be fulfilled." But the dependence on scriptures appears to be even greater than is declared. Randel Helms, a professor of English at Arizona State University, has discussed this in his readable and enlightening book, *Gospel Fictions.*[23]

The evangelists and their sources believed that Jesus was the messiah predicted by the scriptures. They thought the scriptures foretold Jesus' life. *Like many modern believers, the evangelists trusted the scriptures more than any other sources.* Consequently, they believed they could fill in missing facts about Jesus by finding places where these things seemed to be foretold in the scriptures. Thus, they felt justified in using the scriptures to create stories about Jesus.

Our earliest information about Jesus comes from two sources: Q and Paul's letters. According to Mack, both of these sources were written from about 50 C.E. to 80 C.E.[24] As mentioned earlier, Q tells us many of the sayings of Jesus, but says little about his life.

Paul was more involved in passing on his own revelations than furnishing reports of eyewitnesses to Jesus' life. Perhaps Paul *had* little detailed knowledge about Jesus. He proclaimed that Christ died for our sins, was buried, rose after three days, and appeared to Peter, "the twelve," and others; and that all of this was done in accordance with the scriptures. This proclamation, or *kerygma*, may have been based on Isaiah 53 and Ps. 16:10. It also fits the common mystery-cult theme of a god coming to earth, saving humanity, dying, rising, and ascending to heaven.

In *Gospel Fictions*, Helms maintains that Mark, when he wrote the first Gospel in about 70 C.E., did not know much about the life of Jesus. Mark produced a biography of Jesus by collecting, arranging, and linking a large number of mostly fictitious stories. Likewise, the evangelists who wrote the later Gospels probably had little factual information to present. The Gospels of Matthew, Luke, and John were partly based on Mark, but they were written to improve Mark's Gospel.

Improvement was possible, since the original ending of Mark's Gospel was weak, at least for those who wanted abundant proof of the Resurrection. Mark did not include many of the moral teachings of Jesus, so Luke and Matthew brought in these teachings from Q. Mark did not include an *infancy narrative*, but later Gospels provided genealogies and stories about a miraculous conception, annunciations, omens, praise, and persecution of the infant Jesus. Because these parts of the Bible were composed independently, since they could not be based on Mark, they vary greatly and tend to contradict each other.

Helms discusses how *more than thirty miracle stories in the Gospels are based on Old Testament passages*. He concludes that the evangelists or their sources used Old Testament stories as the sources for their stories about Jesus. In many cases, passages that were clearly not presented as prophecies were used to construct Jesus stories. The Jesus stories frequently contain original phrases from the Septuagint, a Greek translation of Jewish scriptures.

In the New Testament, Jesus brought the following dead people back to life: Lazarus (in John), Jairus's daughter (in Mark, Luke, and Matthew), and the widow's son (Luke). In the Old Testament's Books of Kings, Elijah and Elisha each brought a dead person back to life. Helms claims that *"all* the Gospel's stories of Jesus' resurrecting a dead loved one are based on the resurrections in the Books of Kings."[25]

For example, Helms showed how Luke's story of the widow's son matches Elijah's resurrection of the widow's son in 1 Kings 17. Helms points to many parallels between the two stories, such as that both Elijah and Jesus met the widows at "the gate of the city." This particular detail is very peculiar. Archaeological studies of Nain in Galilee, where the New Testament story is placed, showed that the town never had a wall, so there was no gate. Instead, *Nain's fictional gate is in the story because it was part of the Old Testament story.*

The cure of the centurion's servant was reported by Matthew (8:5–13) and Luke (7:1–10). The corresponding Old Testament story is the cure of a woman's dead son by Elisha (2 Kings 4:18–37). In the Old Testament story, Elisha attempted a long-distance cure that failed. Then he went to the dead boy in person and brought him to life. In the Gospel story, Jesus succeeded in curing the centurion's servant, who was sick and near death, *by long distance.* As in other cases, Jesus was portrayed as a greater miracle worker than the ancient prophets.

Another type of miracle involved feeding the multitudes. Helms notes that Mark described *two* miracles by Jesus (6:36–44 and 8:1–21) that are based on the same Old Testament story. In 2 Kings 4:42–44, Elisha fed a hundred people with just twenty loaves and some grain, with food left over. Okay, so maybe the hundred people were just being polite. This doesn't seem likely in the two Gospel stories, however. In the first, five thousand were fed with five loaves and two fishes. In the second, four thousand were fed with seven loaves and a few small fishes. Again, the miracles attributed to Jesus exceeded Elisha's.

Psalm 107 proclaims Yahweh's power over the sea. The psalm starts with God producing a storm on the sea. People in an endangered ship call on the Lord for help. The storm then subsides, and the people are led to their destination. This story was revived and modified in a related story about the prophet Jonah (Jon. 1:3–16). Jonah was regarded as prefiguring the career of Jesus, and Jesus also calmed the sea in Mark 4:35–41. Matthew "improved" on Mark's version by writing it with language closer to that of the Septuagint sources.

Helms maintains that another sea story about Jesus is partly based on the same psalm. This is the story about Jesus walking on the sea during a storm. This may have been suggested by Job 9:8, which refers to God's power to walk on the sea. After seeing Jesus walk on the sea, Peter tried to do it (Matt. 14:28–33):

> Peter called to him: "Lord, if it is you, tell me to come to you over the water." "Come," said Jesus. Peter stepped down from the boat, and walked over the water to Jesus. But when he saw the strength of the gale he was seized with fear; and beginning to sink, he cried, "Save me, Lord." Jesus at once reached out and caught hold of him, and said, "Why did you hesitate? How little faith you have!" They then climbed into the boat; and the wind dropped. And the men in the boat fell at his feet, exclaiming, "Truly you are the Son of God."

Helms says part of this story probably entered Christian folklore from a Buddhist legend. This is possible, since Buddhist missionaries had visited Syria and Egypt by the second century B.C.E. The Buddhist story concerns a disciple who wanted to cross a river to visit Buddha one evening, but found that the ferry was gone.

> In faithful trust in Buddha he stepped into the water and went as if on dry land to the very middle of the stream. Then he came out of his contented meditation on Buddha in which he had lost himself, and saw the waves and was frightened, and his feet began to sink. But he forced himself to become wrapt in his meditation again and by its power he reached the far bank safely and reached his master.[26]

At least one other miracle story may have been influenced by pagan beliefs. Helms discusses John's story of the changing of water into wine at the wedding of Cana. Part of the story's wording matches the story of Elijah (1 Kings 17:8–16) feeding a woman's household and himself for many days from a barrel that initially held only a handful of flour. But that story didn't involve changing water into wine. What was the source of that idea? The idea probably came from a common pagan belief that Dionysus, the Greek god of wine, changed water into wine on his feast day, when certain temple springs were supposed to produce wine instead of water. Following the usual pattern, Jesus' miracle produced much more wine than was involved in Dionysus's miracles.

JESUS' SUFFERING, DEATH, AND RESURRECTION

How literally should we take the Gospel accounts of the suffering, death, and resurrection? They describe many supernatural occurrences. Since such events are not part of our normal experience, reasonable people require overwhelming evidence before accepting such stories as factual. The fact that the stories are recorded in our sacred scriptures is not, in itself, proof that they are accurate reports of historical events. Note that we find it easy to reject the supernatural events in the sacred scriptures of *other* religions. Can we overcome our cultural biases and examine *our own* scriptures in an impartial fashion?

To his followers, Jesus' death was a disaster. Such a dishonorable death suggested that God did not approve of Jesus' work. At first, this was an enormous problem for those who wanted to believe in Jesus. Paul called it the "stumbling block" of the cross. What was needed was evidence that Jesus' death was part of a greater plan.

Almost every person in that part of the world knew a good solution for such a problem. A number of mystery cults were based on the story of a savior coming from heaven to earth, doing great works, being killed, and then *miraculously rising from the dead and returning to heaven*. From a Jewish perspective, if Jesus rose from the dead, this would show God's approval of his life and work. This would be an even greater miracle than Jonah's survival in the belly of a whale or Daniel's survival in the lion's den. It would provide the happy ending also found in other Old Testament stories such as those about Esther, Joseph, and Susanna. Such a solution appealed to both gentiles and Jews. *The situation demanded that Jesus rise from the dead!*

But what was the *evidence* that Jesus had risen from the dead? Before too many years after the death of Jesus, two forms of evidence appeared. One was the story of the empty tomb on Easter morning, as reported by Mark and copied, with some changes, in later Gospels. The second type of evidence was the stories of appearances of Jesus after his alleged resurrection. *Both kinds of evidence were only stories*, and we need to keep in mind that *the stories came from people who were not actual eyewitnesses to the events in the stories, and these people wanted to spread the Christian message.*

Concerning their origins, the accounts of the suffering, death, and resurrection seem to separate naturally into three parts. The first part is the *arrest, trial, suffering, and death of Jesus*. This part is largely based on the Jewish scriptures. The second and third parts are the stories of *the empty tomb* and *the appearances of Jesus after the Resurrection*. To allow Christianity to spread among Jews and gentiles, these parts were needed in order to make Jesus' death a glorious event, rather than just another human tragedy. The resurrection story was unprecedented in Jewish experience, so few scriptures suggested the details found in those stories.

Let's discuss the first part, from the arrest to the death of Jesus. Bible scholars in the Jesus Seminar have attempted an impartial evaluation of this story. The seminar's founder, Robert Funk, has concluded that

> The story of Jesus' arrest, trials, and execution is largely fictional. It was based on a few historical reminiscences augmented by scenes and details suggested by prophetic texts and the Psalms.[27]

Also concerning this story, called the passion narrative, Burton Mack says,

> The usual approach to Mark's so-called passion narrative has
> been to regard it as a historical account of what really happened,
> but then to fret about features of it that are difficult to accept. . . .
> The better approach is to recognize the whole story as Mark's fic-
> tion, written forty years after Jesus' time in the wake of the
> Roman-Jewish war.[28]

John Dominic Crossan is a cofounder of the Jesus Seminar and
a professor of biblical studies at DePaul University in Chicago. In
Who Killed Jesus? his recent and very illuminating book, he
describes the passion narrative as 20 percent "history remembered"
and 80 percent "prophecy historicized."[29] By *history remembered*,
Crossan means parts of the story that were actually observed or
reliably inferred by people who contributed to the Gospel tradition.
By *prophecy historicized*, he refers to the process described in the
previous section, in which Christian scribes searched the Jewish
scriptures for supposed predictions about Jesus, and then recorded
these items as if they were actual events in the life of Jesus.

What parts of the passion narrative belong in the 20 percent
that Crossan believes were actual historical events? He believes that
Jesus was a real person and that he was executed under Pontius
Pilate. These facts are supported by Flavius Josephus and Cornelius
Tacitus, two ancient historians who were not biased in favor of the
Christian message. Both mention that the Christian movement continued after Jesus' death, and Tacitus reports that it had even
spread to Rome, "where all things horrible or shameful in the world
collect and find a vogue."

After Jesus was arrested, it is likely that the disciples ran away.
If they did, they would not have been witnesses to the scourging,
mocking, and crucifixion of Jesus. Crossan reports, however, that
the Romans frequently used crucifixion to execute people. It was
also standard procedure to beat or whip condemned people before
they were crucified. Thus, Jesus' scourging and crucifixion would
have been fairly predictable, even if they were not witnessed.

Helmut Koester, a professor of New Testament and ecclesiastical history at the Harvard Divinity School, says,

> One can assume that the only historical information about Jesus' suffering, crucifixion, and death was that he was condemned to death and crucified. The details and individual scenes do not rest on historical memory, but were developed on the basis of allegorical interpretations of Scripture.[30]

In a table, Mack gives thirty-nine features of the passion narrative, with their references in Mark as well as in their likely Old Testament sources. Of the thirty-nine, most come from the psalms, with at least five references to Psalm 22. There are still other parallels, such as the "Darkness at Noon" episode during the crucifixion, reported by Mark, Matthew, and Luke, and which Crossan traces back to Amos 8:9–10.

Crossan also discusses a complicated relationship between the passion narrative and the Jewish scapegoat ritual. In the ritual, the people would symbolically transfer their sins to a goat that would be taken away into the wilderness. Crossan notes many parallels between the ritual and the passion narrative. Clearly, Jewish scriptures and traditions served as a quarry for extracting details of the passion narrative!

Although it is not in the Bible, another Gospel presented a passion narrative. It is the *Gospel of Peter*. It is only partly known to us from a few fragments, but what it contains is very interesting. It can be found in an appendix to Crossan's *Who Killed Jesus?* Some scholars think it was an earlier version of the four Gospels. Compared with them, its wording is closer to the Old Testament scriptures. According to Robert J. Miller, a member of the Jesus Seminar, "almost every sentence of the passion narrative of Peter appears to be composed out of references and allusions to the psalms and the prophets."[31] The Gospel of Peter is supporting evidence that the passion narratives in the Gospels were created by using ancient scriptures to compose what would later be taken as a historical account of Jesus' final days.

Now let's discuss the empty-tomb stories. These are part of the evidence for the resurrection of Jesus. It was mentioned above how important it was for the early Christian movement that Jesus actually rose from the dead. Paul even wrote that their faith was in vain if Jesus did not rise from the dead (1 Cor. 15:14). But Paul never mentioned an empty-tomb story. Surely Paul would have used such a story if it existed at the time of his writings and if he knew it. Instead, Paul merely *proclaimed* that Jesus had risen from the dead and had appeared to certain followers.

The earliest empty-tomb story in the Gospels is that of Mark. It goes as follows:

> When the sabbath was over, Mary of Magdala, Mary the mother of James, and Salome bought aromatic oils intending to go and anoint him; and very early on Sunday morning, just after sunrise, they came to the tomb. They were wondering among themselves who would roll away the stone for them from the entrance to the tomb, when they looked up and saw that the stone, huge as it was, had been rolled back already. They went into the tomb, where they saw a youth sitting on the right-hand side, wearing a white robe; and they were dumbfounded. But he said to them, "Fear nothing; you are looking for Jesus of Nazareth, who was crucified. He has been raised again; he is not here; look, there is the place where they laid him. But go and give this message to his disciples and Peter. 'He is going on before you into Galilee; there you will see him, as he told you.'" Then they went out and ran away from the tomb, beside themselves with terror. They said nothing to anybody, for they were afraid. (Mark 16:1–8)

Both Funk and Crossan maintain that the empty-tomb story is fictional and was probably made up by Mark. Funk says,

> The Fellows of the Jesus Seminar concede that Jesus may possibly have been buried in a common grave, but they doubt that his grave site was ever known. . . . The story of the women discovering an empty tomb on Easter morning, which suggests that a burial site was known, was undoubtedly a literary creation of Mark.[32]

Crossan says,

> I do not find anything historical in the finding of the empty tomb, which was most likely created by Mark himself—at least I cannot find it anywhere except under his influence.[33]

By "under his influence," Crossan means that Matthew, Luke, and John based their versions of the empty-tomb story on Mark's story. However, these evangelists did not preserve the story just as Mark had written it. Consider how they describe who went to the tomb and how many angels were seen.

Mark: Two Marys and Salome went to the tomb and saw one angel.
Matthew: Two Marys went to the tomb and saw one angel.
Luke: Two Marys, Joanna, and others went to the tomb and saw two angels.
John: Mary Magdala went to the tomb and saw two angels.

In these and other details, the later evangelists changed Mark's story. *They treated it as if it were fictitious and could be improved at will.* Obviously, none of them imagined that four different Gospels would be bound together in one book, so that the discrepancies would be so evident.

The appearances of Jesus after the Resurrection make up the final part of the passion and resurrection narratives. Unlike his silence concerning the empty-tomb story, Paul does mention some appearances of the risen Christ. In 1 Cor. 15:4–8, Paul declares that Jesus:

> was raised on the third day, in accordance with the scriptures, and that he appeared to Cephas, then to the twelve. Then he appeared to more than five hundred brothers and sisters at one time, most of whom are still alive, although some have died. Then he appeared to James, then to all the apostles. Last of all, as one untimely born, he appeared also to me.

Paul seems to refer to Cephas (Peter) and the twelve as other than James and the apostles, but let's not sidetrack the discussion. Jesus' famous appearance to Paul on the road to Damascus is treated by Paul as if it were equivalent to the other appearances, although it presumably was a few years later. This appearance was described three times by Luke in the Acts of the Apostles (9:3–4, 22:6–7, and 26:13–14).

> Now as he was going along and approaching Damascus, suddenly a light from heaven flashed around him. He fell to the ground and heard a voice saying to him, "Saul, Saul, why do you persecute me?" (Acts 9:3–4)

For many modern readers, this story suggests that Paul had experienced some kind of seizure or unusual state of consciousness rather than a supernatural event.

We might expect to see the appearance stories next in Mark, but this Gospel originally ended at Mark 16:8, with the quote given a few pages back: "Then they went out and ran away from the tomb, beside themselves with terror. They said nothing to anybody, for they were afraid." Helms discusses this as follows:

> The most ancient manuscripts of Mark end at this point, one of the strangest and most unsatisfying moments in all the Bible, depicting fear and silence on Easter morning and lacking a resurrection appearance. But within about fifty years, at least five separate attempts were made by various Christian imaginations to rewrite Mark's bare and disappointing story; they appear in the Long Ending and the Short Ending of Mark, and in the Gospels of Matthew, Luke, and John. The first two are second-century interpolations in some texts of Mark and are identified as such in any responsible modern text. They are Mark 16:9–20 (in the King James Version and others based on late manuscripts), an unskillful paraphrase of resurrection appearances in other Gospels; and Mark 16:9 in a few other late manuscripts, in which the women followed the youth's instructions to tell the disciples, a statement that conflicts with verse 8 of the original text.[34]

Matthew, Luke, and John were not able to copy appearance stories from Mark, since there were none. Mark must not have known any appearance stories, or, for some reason, he chose not to include them in his Gospel. The whole passion narrative, however, cried out for appearance stories to give it a more satisfactory ending. Thus, all three of the later evangelists collected or composed such stories and added them to their Gospels. Since they were probably not in communication with each other, their appearance stories are not consistent with each other. Here is a summary of the appearances recorded by Paul and the three later evangelists.

- *Paul* had Jesus appear to Peter, to the twelve, to more than five hundred followers at once, to James, to all the apostles, and finally to Paul himself.
- *Matthew* had Jesus appear to Mary Magdala and another Mary near the tomb, and to the eleven apostles (Judas was dead) on a mountain *in Galilee.*
- *Luke* had Jesus appear to two disciples near Jerusalem, to Peter, and, a little later, to the eleven *in Jerusalem,* after which Jesus led the disciples a short distance to Bethany, where he rose up into heaven.
- *John* has Jesus appear to Mary Magdala at the tomb, then to the disciples *in Jerusalem,* and eight days later to the disciples again. Later, according to John, Jesus appeared to the disciples at the sea of Tiberias, *in Galilee.*

The stories list plausible people and locations for the appearances, but they have little in common. The appearance stories are about as different as they might be expected to be if the evangelists recorded different samples drawn from a vast body of mythical traditions about Jesus.

Some people may actually have had some visions of Jesus after his death. His followers probably prayed, fasted, and thought continually about him. Lengthy periods of fasting and praying may have led to trances or visions, just as fasting and praying led to visions in the traditional vision quests of young members of certain Native American tribes.

Visions of Jesus may have been the basis for stories of his appearances after his death, but it proves nothing if such visions occurred.

That people can see Jesus when they are in a trance or ecstasy does not prove the validity of the Christian message. This could happen in any religion, but, by itself, it would not prove that the religion sponsoring this activity has any special connection with reality.

The three Gospels that originally described the appearances of Jesus implied that he appeared *in the flesh*, not just by ghostly apparitions, visions, or hallucinations. In Matthew, the two Marys worshiped Jesus and *held him by the feet* after he appeared. In Luke, Jesus appeared to the eleven and told them he was not a spirit, he showed his wounds, and ate fish and honeycomb. Similarly, in John, Jesus showed his wounds, breathed on the disciples, invited the "doubting Thomas" to put his hand into the wound in his side. Later, Jesus presumably ate bread and fish with the disciples at the sea of Tiberias. Luke and John make a point of stressing the belief that Jesus' appearances were physical.

Why did the Gospels stress the doctrine that Jesus appeared with a physical body? One possible answer is that this made the stories much more impressive, since mere apparitions seemed to occur in many other, presumably "false," religions.

Funk maintains that the *uniqueness* of Jesus' physical resurrection served as a foundation for the authority of the early church.[35] During his appearance at the sea of Tiberias, Jesus commanded Peter to "Feed my lambs" and "Feed my sheep." In Luke 24:47 Jesus laid out a plan for the eleven to preach a doctrine of repentence and the remission of sins to all nations, beginning in Jerusalem. Thus, these bodily appearances of the risen Jesus *gave authority to Peter and the eleven to carry out the work of the church*. In this view, later visions *that were not physical*, such as those of Paul and, perhaps, many others, did not bestow the majestic powers that had been granted to Peter and the apostles by the risen Jesus.

We need to keep in mind that, although these things were written in the Bible, we have no convincing evidence that they were not entirely fictitious. It is likely that *most* of the Gospel material is fictitious, except for some sayings of Jesus in Q[1] and a few basic facts about his life.

In summary, scholars believe that the Jesus movement

upgraded the sayings of Jesus until he presented himself as a divine personage. The Christ cult fitted Jesus' life into the mythical pattern of a god who came to earth to save humanity, was killed, rose from the dead, and returned to heaven. The framework of the Gospel accounts of Jesus' life was invented by Mark and copied by the other evangelists. Numerous well-meaning believers contributed their "revelations" to evolving stories about miracles, a resurrection, and after-death appearances. Many of the revelations were not freely created fictions, but were based on Old Testament stories. Often these stories had not originally been presented as prophecies.

Skepticism about the Gospels leads to skepticism about other supernatural beliefs that are, in practice, connected with faith in the Bible. It appears that the Bible originated by entirely natural processes and that its supernatural events are fictitious. As a result, we have little reason to trust other ideas about supernatural events and entities that are loosely linked to the Bible. So, we are prepared to investigate whether the soul exists, without worrying about the fact that it is *assumed* to exist in passages in the Bible.

NOTES

1. Burton L. Mack, *The Lost Gospel: The Book of Q and Christian Origins* (San Francisco: HarperSanFrancisco, 1993), p. 24.

2. Burton L. Mack, *Who Wrote the New Testament? The Making of the Christian Myth* (San Francisco: HarperSanFrancisco, 1995), pp. 62, 63.

3. Arthur G. Patzia, *The Making of the New Testament* (Downers Grove, Ill.: InterVarsity Press, 1995), p. 55.

4. Mack, *The Lost Gospel* and *Who Wrote the New Testament?*

5. Robert W. Funk, *Honest to Jesus: Jesus for a New Millennium* (San Francisco: HarperSanFrancisco, Polebridge Press, 1996).

6. Mack, *The Lost Gospel*, pp. 81–102.

7. John S. Kloppenborg, *The Formation of Q, Trajectories in Ancient Wisdom Collections* (Philadelphia: Fortress Press, 1987), pp. 243–45.

8. Mack, *The Lost Gospel*, p. 37.

9. Ibid., p. 247.

10. Ibid., pp. 159–62.

11. Ibid., p. 48.

12. Ibid., p. 54.

13. Ibid., pp. 114–21.

14. Funk, *Honest to Jesus*, p. 79.

15. Mack, *The Lost Gospel*, pp. 191–205.

16. Maurice Casey, *From Jewish Prophet to Gentile God: The Origins and Development of New Testament Christology* (Louisville, Ky.: Westminster/John Knox Press, 1991), p. 27.

17. Mack, *The Lost Gospel*, p. 192.

18. Mack, *Who Wrote the New Testament?* p. 80.

19. Ibid., p. 153.

20. Funk, *Honest to Jesus*, pp. 281, 282.

21. Ibid., pp. 282–94.

22. Esmé Wynne-Tyson, *Mithras: The Fellow in the Cap* (New York: Barnes and Noble, 1972), p. 83.

23. Randel Helms, *Gospel Fictions* (Amherst, N.Y.: Prometheus Books, 1988), pp. 19, 20, 35–147.

24. Mack, *The Lost Gospel*, Appendix A.

25. Helms, *Gospel Fictions*, p. 65.

26. Ibid., p. 81.

27. Funk, *Honest to Jesus*, p. 127.

28. Mack, *The Lost Gospel*, p. 158.

29. John Dominic Crossan, *Who Killed Jesus?* (San Francisco: HarperSanFrancisco, 1995), p. 1.

30. Helmut Koester, *Ancient Christian Gospels: Their History and Development* (Philadelphia: Trinity Press International, 1990), p. 224.

31. Robert J. Miller, *The Complete Gospels* (San Francisco: HarperSanFrancisco, 1994), p. 400.

32. Funk, *Honest to Jesus*, pp. 220, 221.

33. Crossan, *Who Killed Jesus?* p. 209.

34. Helms, *Gospel Fictions*, p. 133.

35. Funk, *Honest to Jesus*, pp. 271–73.

Part II

SCIENCE AND THE SOUL

5

THE BLOCKS AND CEMENT
THAT BUILD OUR WORLD

I N LATER CHAPTERS, I WILL present a worldview that is more scientific and less traditional than the views held by most people. A reasonable person, however, does not change his or her worldview without obtaining a new understanding of how the world works. To allow readers to reach this new understanding, I will use this and the following two chapters to describe how science explains much of what happens in our world.

These chapters may seem far removed from the discussion of the issues of free will, consciousness, and the soul. This is not the case, however, since the success of science in understanding nature is at the heart of our later discussion of those issues. It is difficult to appreciate the arguments given later without some idea of how science has succeeded in explaining so much about the world.

Our world operates by means of orderly interactions of the basic building blocks of nature. Science has not uncovered any exception to this rule. The building blocks, called *elementary particles*, are described in this chapter. They are important because of their role in forming atoms and because they were involved in the origin of matter, discussed in the next chapter. Some particles that are even more basic than the so-called elementary particles will also be described. These are the *quarks* and *gluons*. Thus, this chapter deals with three levels: atoms, elementary particles which combine to make atoms, and the quarks and gluons which combine to make elementary particles.

The types of forces that act on these particles will also be described. There are four basic forces in nature, and they form the "cement" which binds the building blocks together. These forces between interacting particles result from the continual interchange of other particles between the interacting particles.

No special background in science is needed to understand this chapter and the following chapters. It is necessary, however, to introduce one technique that is helpful in writing very small and very large numbers. This is writing numbers by using scientific notation. A large number is changed into a first number multiplied by a second number. The first number is between 1 and 10. The second number is 10 multiplied by itself a number of times. As a short way of writing this, the second number is represented by 10 followed by a superscript which is the number of tens which multiply each other.

Let's see how scientific notation works by looking at a few examples. Consider the number 65,500. We can rewrite it as $6.55 \times 10 \times 10 \times 10 \times 10$ or 6.55×10^4. We can rewrite 5,300,000 as $5.3 \times 10 \times 10 \times 10 \times 10 \times 10 \times 10$ or just 5.3×10^6. Similarly, when we have a number that is smaller than 1, we can put it in scientific notation by dividing by 10 a certain number of times. Since dividing by 10 is the opposite of multiplying by 10, we use a negative superscript in this case. For example, 0.000037 is 3.7 divided by 100,000 (which is 10^5) so it is just 3.7×10^{-5}.

THE ATOM IN CLASSICAL GREECE AND MODERN SCIENCE

Compared with the scale of the basic building blocks of nature, we human beings are enormous. Since the building blocks are so small, our ordinary experience suggests that all materials can be divided into arbitrarily small pieces. Science has shown, however, that one cannot divide matter into pieces smaller than a certain size without radically changing the properties of the material. These minimum-sized packages are called *atoms*. Atoms are of interest for their own sake, for their importance in explaining so much in chemistry, and for their role as a natural introduction to the elementary particles found in nature.

The idea of atoms occurred long before it was possible to prove that they exist. In the fifth century B.C.E. the Greek philosophers Leucippus of Miletus and Democritus of Abdera proposed that matter is made of atoms, which move through empty space. It was believed to be impossible to break these atoms into smaller pieces.

The atoms, according to Democritus, interact with each other in a mechanical way. This model of the world implied that the interactions of the atoms determined all movements and changes, even on the large scale of humans.[1] In the following centuries, many philosophers and Jewish and Christian teachers rejected this picture, perhaps because it seemed to limit personal moral responsibility for one's actions. Also, there was no convincing evidence supporting this picture. Consequently the views of the "atomists" were set aside by most thinkers of early Christianity and the Middle Ages, but the ideas were to become very important later.

Two enormously important contributors to modern science, Galileo (1564–1642) and Isaac Newton (1642–1727), believed in atoms,[2] but they did not find evidence to prove that atoms exist. Much of the early evidence in support of the atomic theory came from chemistry. Mikhail Vasilyevich Lomonosoff (1711–1765), in 1756, and Antoine Laurent Lavoisier (1743–1794), in 1774, stated the law of conservation of mass in chemical reactions. The law is that the total weight (amount of mass or matter) of the substances formed in a chemical reaction is equal to the total weight of the substances that enter into the reaction. This law helped to lead chemistry toward the discovery of the atomic theory.

Another important step toward the atomic theory was the identification of the chemical *elements*. Elements, such as iron, carbon, and oxygen, are pure substances that cannot be separated into other more basic constituents by chemical processes. Much work by many workers was involved in identifying the elements. Many of the elements were listed in a "Table of Simple Substances" published in Lavoisier's *Traité Elémentaire de la Chimie* in 1789.

By mixing and heating samples of different elements, chemists produced chemical compounds. Careful weighing of the materials *actually used in these chemical reactions* showed that each chemical compound requires a specific ratio of weights of its different ingredients. In a few cases two different compounds could be formed from the same two elements. In these cases, for a certain weight of one element, the weights of the other element used to make the two compounds were, amazingly, exactly in the ratios of small whole numbers, like 1 to 2, 1 to 3, or 3 to 4.

These regularities led John Dalton (1766–1844) to propose his atomic theory in 1808. His theory explained the facts mentioned above. Dalton's theory used the following assumptions:

1. All matter consists of minute, discrete particles.
2. In *elements* these particles are indivisible, indestructible *atoms* that are not changed by physical or chemical processes.
3. All atoms of a given element are alike in all ways, including weight, and they differ from the atoms of all other elements.
4. In compounds, the minute particles are called *molecules*. Compounds are formed from elements by the combination of atoms to form molecules. Atoms combine in simple numerical ratios such as 2 to 1, 1 to 2, and 2 to 3.

Evidence for the existence of atoms also came from the properties of gases. Measurements of pressure, volume, and temperature give simple results for all gases under conditions where the gases are far from forming a liquid. A sample of any gas obeys the law that *the product of the gas's pressure and volume, divided by the temperature, gives a number that stays the same as the pressure, volume, and temperature change.* To apply this law, pressure must be measured on a scale for which zero is the pressure of a vacuum. Similarly, temperature must be measured on a scale in which zero is the lowest possible temperature.

The law given above and the nature of heat in gases were explained by the *kinetic theory of gases* developed by several scientists between 1738 and 1900. The theory involves atoms and molecules. In this theory, the pressure of a gas on a container's walls results from the bombardment of the walls by the atoms or molecules of the gas. It was shown that the product of the pressure and the volume is proportional to the total energy of motion, or *kinetic energy*, of all the atoms or molecules in the gas. The kinetic energy is proportional to the temperature of the gas. This is all that was needed to prove that the law mentioned above should hold.

The kinetic theory of gases also gave correct predictions for how the temperature of a gas changes when heat is added. In addition, the theory predicted the distribution of speeds of gas molecules. The speeds of gas molecules were measured and the results agreed with the predicted distribution. The great predictive success of this theory added to the support for the underlying atomic theory.

WEIGHING THE ATOM

Despite the usefulness of the atomic theory in describing so many facts, it was still possible to doubt the reality of atoms as late as 1900. No one had seen atoms because they are too small to be seen with a microscope. The closest thing to seeing them was to use a microscope and look at tiny pollen grains suspended in water. The pollen grains jump around continually due to their constant bombardment by rapidly moving atoms or molecules!

No way had been found to find the weight of the atoms. Dalton's atomic theory and chemical experiments were used to find the *relative weights of different types of atoms*. Hydrogen has the lightest atoms. A carbon atom is about twelve times heavier than a hydrogen atom, and an oxygen atom is about sixteen times as heavy as a hydrogen atom. But no one knew how many hydrogen atoms it takes to make a gram of hydrogen. The answer to this question is approximately equal to Avogadro's number, named after the Italian physicist Amedeo Avogadro (1776–1856).

The French physicist Jean Perrin (1870–1942) measured Avogadro's number in 1908. A suspension of very tiny spheres of gum resin was held in a liquid drop. These tiny balls would settle to the bottom of the drop if they were not continually bombarded by the movement of atoms in the fluid. The bombardment kept the balls off the bottom, but they were not uniformly distributed throughout the droplet. The highest density was near the bottom and the density decreased continually at greater heights above the bottom. The decrease of density with increasing height was measured and was used, after some algebra, to calculate Avogadro's number. The number was about 6×10^{23}. This means that there are about 6×10^{23} atoms in a gram of hydrogen!

The mass of a hydrogen atom was calculated from the number of atoms it takes to make a gram of hydrogen and was found to be about 1.7×10^{-24} grams. Since the relative weights of other atoms were known, their masses could also be calculated. Perrin, in effect, was the first person to "weigh" the atom. Weighing the atoms gave strong evidence that atoms are real.

We can make a rough estimate of the *sizes* of atoms using Avogadro's number, the known atomic weights relative to hydrogen, and the measured densities of solids. The computed sizes are typically 2×10^{-8} to 3×10^{-8} cm and agree with the sizes obtained by more complicated methods. (One centimeter or "cm" is 0.39 inches.)

ELEMENTARY PARTICLES WITHIN THE ATOM

Some of the most important and basic studies of the nature of the physical world involved breaking atoms apart and examining the pieces. According to Democritus and Dalton this should be impossible, since atoms were assumed to have no parts. Certain experiments involving electrical phenomena led researchers to conclude that *atoms do have parts,* which we now call *subatomic or elementary particles.* These are more fundamental building blocks of matter than the atoms. Before going on, some simple knowledge of electricity is needed.

We are all familiar with static electricity causing clothing or pieces of paper to "cling" to each other. Rubbing a hard rubber rod with fur, or rubbing a glass rod with silk, leaves an electrical charge on the rubber or glass. A pair of very light balls, called *pith balls,* can be suspended on strings so that they hang near each other. If both balls are touched by the hard rubber rod some of the charge is transferred to the balls, and the balls repel each other. If both are touched by the glass rod they also repel each other. If one pith ball is touched by the rubber rod and one is touched by the glass rod, however, the balls attract each other.

Other experiments show that an object that has been charged by a rubber rod can have the charge canceled by being touched by a charged glass rod. The conclusion of all this is that *there are two opposite kinds of electrical charge.* By convention, the charge on the glass rod is called positive and the one on the hard rubber rod is negative. (This convention was established by Benjamin Franklin [1706–1790], who developed an early theory of electricity.) The experiments with the pith balls show that like charges repel each other and unlike charges attract each other.

Other experiments show that matter is normally electrically neutral; that is, it has no net electrical charge. The rubber rod and the fur are initially neutral. When the rod is rubbed with the fur, negative charge is transferred from the fur to the rubber rod. The fur is left with a positive charge of the same amount as the negative charge on the rubber rod. The conclusion is that electrical charge is a conserved quantity. That is, starting with uncharged objects, when one thing receives a negative charge there is an equal amount of positive charge produced on something else. The process of generating static charges on objects does not change the net amount of charge that is present on all of the bodies that are involved in the process.

The principle that charge is conserved is an example of a *conservation law*. Conservation laws are very important in nature. We discussed the conservation of mass in chemical interactions. Another conservation law is the conservation of *energy*. With two exceptions (see below), the total amount of energy is unchanged in all processes. Energy can be transformed from one form to another, as when a machine loses kinetic energy by friction and produces heat. Energy has many forms, such as kinetic energy, sound, light, heat, and potential energy. (Potential energy is stored energy, as in a wound-up spring in a clock or a boulder on top of a cliff.)

One exception to the law of conservation of energy is that energy can sometimes be converted into matter, and matter can sometimes be turned into energy. This process obeys Einstein's famous equation that $E = mc^2$, which says that an enormous amount of energy can be converted into a tiny amount of mass, or vice versa. This convertibility between mass and energy is included in an improved conservation law, the law of *conservation of mass-energy*. This says that the sum of E (energy other than mass) and mc^2 (mass times the speed of light squared) is always conserved.

Another exception to the conservation of energy is allowed by *quantum mechanics*, the theory that applies to what happens to extremely small objects, such as elementary particles and atoms. In this case energy conservation can be violated by extremely small amounts for extremely short intervals of time.

Now let's return to the evidence that atoms have parts. The first experiments showing that atoms have parts used gas-discharge tubes. These are closed glass tubes with metal plates at each end. The two plates were charged to a high voltage (about 600 volts). The tubes were connected to a vacuum pump and the air pressure was reduced to about 0.5 percent of normal atmospheric pressure. Under these conditions the low pressure gas emitted a purple glow, if the tube had initially been filled with air. (If neon gas had been in the tube a red glow would have been visible, as you have seen many times in advertising signs.)

By putting an object into the gas discharge tubes it was found that the glow seemed to be produced by a stream of particles called *cathode rays.* The rays traveled in straight lines and shadows (absence of glowing) were produced on the side of the object away from the negatively charged metal plate (called the *cathode*). This implied that the rays were moving away from the cathode, as if they carried negative charge and were repelled by the negative charge on the plate. Magnetic fields deflected these rays, confirming that the particles carried negative electic charge. Small paddlewheels were put into the tubes so that one side was struck by the rays. The paddlewheels rotated under the impact of the rays. This showed that the rays carried momentum.

Two experiments were crucial in understanding the nature of cathode rays. One was done by J. J. Thomson (1856–1940) at Cambridge University in 1897. He used a cathode-ray beam in a low-pressure discharge tube. He determined the speed of the cathode rays by finding the precise strength of a magnetic field which canceled the deflection of the beam produced by an electric field. Using this speed as well as the deflection produced by the electric field alone, he calculated the ratio of the charge to the mass of these rays. This quantity, the charge-to-mass-ratio, was the same for all pressures, all kinds of gas, and all kinds of metal plates in the discharge tubes. Since their properties don't depend on the experimental details, this led to the idea that the rays are one of the most basic particles in nature, called *electrons*. Later experiments supported this idea.

The second crucial experiment with electrons was done by R. A. Millikan (1868–1953) at the University of Chicago about 1911. He used a microscope to observe very tiny oil droplets moving in air. The oil droplets were affected by four forces. Gravity and the buoyancy of the air always affected them. Friction with the air tended to oppose the motion of the droplets. Some oil drops had small electric charges, and Millikan used electrically charged plates above and below the oil drops to produce a fourth force on those drops.

By many observations of droplets, with the metal plates both charged and uncharged, and by careful analysis of the forces on the droplets, Millikan measured the charges on the droplets. This charge always had certain values that were almost exactly one, two, three, or more times a specific charge, called e. That is, the droplets always carried a whole number of charges of size e, and e was then taken to be the charge carried by every electron. The droplets with one, two, three, or more times the electron's charge were carrying one, two, three, or more extra electrons, respectively.

At this point the electron's change was known to be $-e$. (The electron's charge is negative, but e represents a positive number, the charge's magnitude.) Since the charge-to-mass-ratio had been measured by Thomson, the electron's mass could be computed. The mass turned out to be 9.1×10^{-28} grams. Since the hydrogen atom's mass is 1.7×10^{-24} grams, the hydrogen atom is almost two thousand times heavier than the electron. The electron is only a part of an atom, and almost all of the mass of the atom must be carried by one or more other particles. Since hydrogen is normally electrically neutral, the other part of the atom must have a positive charge, equal to e. The evidence pointed to other building blocks in nature besides the electron.

RADIOACTIVITY AND THE DISCOVERY OF THE NUCLEUS

Let's take a short detour to discuss radioactivity, which was used in some studies of the atom. Radioactivity was discovered by Antoine Becquerel (1852–1908) in 1896. He found that certain uranium compounds could expose photographic plates wrapped in light-proof materials. The radiation continued for months without major weakening. About 1900, Pierre (1859–1906) and Marie (1867–1934) Curie laboriously extracted tiny samples of two new radioactive elements, polonium and radium, from enormous amounts of pitchblende. (Pitchblende is an ore containing uranium and some other heavy elements.) Polonium and radium are enormously more radioactive than uranium. They served as radiation sources in many later experiments.

Ernest Rutherford (1871–1937) carried out many of the early investigations of radioactivity. He was born in New Zealand, but did his major scientific work in Canada and England. Studies of radioactivity revealed three kinds of radiation, arbitrarily named *alpha rays*, *beta rays*, and *gamma rays*. Rutherford identified the first two kinds in 1897. Alpha rays are the least penetrating and are stopped by a thick sheet of paper. Alpha rays are fast-moving helium *nuclei*. (Nuclei will be discussed later.) Beta rays are rapidly moving electrons. Gamma rays are very penetrating and are a form of electromagnetic radiation, with shorter wavelengths than X-rays.

We saw above that electrons were found within atoms, and that there was evidence for at least one other basic building block of matter within atoms. Rutherford carried out the next important steps in exploring the atom in 1909–1911. With his assistants Geiger and Marsden, he used radium as a source of alpha particles to bombard an extremely thin gold foil. He knew that the beams of alpha particles would not be scattered significantly by the electrons in the atoms. The experiment was done to learn the properties of the positive particles (the *protons*) within atoms. If the positive particles were spread all over the atoms, then the alpha particles would frequently be deflected by a small angle by collisions with one of the many protons in a gold nucleus. The alpha particles would never be deflected by very large angles, however, since they are about four times heavier than protons.

What Rutherford and his collaborators were extremely surprised to observe, however, was that some alpha particles were deflected by very large angles. Describing this result, Rutherford said, "It was quite the most incredible event that has ever happened to me in my life. It was almost as incredible as if you fired a 15-inch shell at a piece of tissue paper and it came back and hit you." This large angle scattering showed that the protons are not spread out within the atom, but are all bound together in a very small and heavy group, called *the nucleus of the atom*. (The plural of nucleus is nuclei.) The very large deflections were extremely rare because the nucleus is only a tiny target within the atom. The diameter of the gold nucleus is about 10^{-12} cm. The gold atom has a diameter about twenty-five thousand times greater than the diameter of the nucleus!

J. J. Thomson, studying positive rays in gas-discharge tubes filled with hydrogen, measured the charge-to-mass ratio of hydrogen nuclei. Thomson correctly assumed that *these nuclei have an electrical charge equal to e* and he found that the hydrogen nucleus has a mass 1,836 times heavier than the electron's mass.

In 1919 Rutherford bombarded nitrogen with alpha rays. He observed that massive particles were sometimes released in these collisions. The mass and charge of these particles matched those of the hydrogen nuclei studied by Thomson. *Rutherford concluded that protons had been generated in the collisions and that protons are the positive particles in the nuclei of atoms.*

In the work described above, Rutherford also observed traces of oxygen in an initially pure sample of nitrogen gas following the bombardment by alpha particles. This proved that nuclei were being changed from one kind to another in this process. In other words, *Rutherford had discovered a process by which one chemical element was changed into another!* Much earlier, alchemists had tried to produce gold out of other materials. Scientists eventually found ways to produce any element from other elements, but at a cost so high that it is almost always cheaper to mine the desired elements. An exception is that certain radioactive materials that are extremely rare in nature are produced by such artificial means. These materials are valuable in certain medical processes, including the treatment of cancer.

In 1932 James Chadwick found a penetrating form of radiation by bombarding beryllium with alpha particles. This radiation involved a neutral particle with a mass almost identical to the proton's mass. These particles were called *neutrons. With the discovery of neutrons we have a complete list of the building blocks that make up atoms.* Neutrons and protons form the nuclei of atoms. These two kinds of particles that are found in nuclei are called *nucleons.* Surrounding the nucleus is a much larger cloud of electrons. The chemical properties of atoms are determined primarily by the number of electrons in the neutral atoms. In a neutral atom, the number of electrons is equal to the number of protons.

All the atoms of an element have the same number of protons

in the nucleus. The number of neutrons in the nuclei of a given element can vary. (This violates assumption 3 of Dalton's atomic theory, but it does not make much difference to the chemistry of the atoms.) Let's consider carbon, for example. Carbon usually has twelve nucleons (six protons and six neutrons). The chemical symbol for carbon is C. The symbol for a nucleus of carbon with twelve nucleons is ^{12}C. There is also a form of carbon with six protons and eight neutrons (^{14}C).

An *isotope* is a material that has a particular number of protons and another particular number of neutrons in its nuclei. So, ^{12}C and ^{14}C are different isotopes of carbon. A lone proton is found in the nucleus of most hydrogen (^{1}H). A small fraction of hydrogen, however, has a nucleus containing a proton and a neutron (^{2}H). Ordinary hydrogen is 99.98 percent ^{1}H and 0.02 percent ^{2}H. A radioactive isotope of hydrogen can be produced artificially which contains one proton and two neutrons in the nucleus.

PARTICLES THAT HOLD THE NUCLEUS TOGETHER

What is the cement that holds an atom together? Electrons are attracted to a nucleus and held in orbit around the nucleus because electrons are negatively charged and the nuclei are positively charged. (Remember that oppositely charged pith balls attract each other.) It is not so easy to explain why the nucleons stay together in a nucleus. (Physics is just one problem after another, isn't it?) Gravity is far too weak to hold the nucleons together. The electrical force between the protons in a nucleus tends to make them repel each other, since all protons have positive charges. Neutrons do not have any charge, so they do not attract or repel other nucleons by electrical forces. A new kind of force was proposed, called the *strong force*. It is strong enough to overwhelm the electrical repulsion between the protons. The strong force is an attraction between any pair of nucleons, whether the nucleons are neutrons or protons.

An early model of the strong force was proposed by the

Japanese physicist Hideki Yukawa (1907–1981) in 1935. In his theory, the attractive force between nucleons is due to the continual exchange between the nucleons of (guess what?) a new kind of particle. The exchanged particle is not permanently present, and the process occurs even though there is not enough energy available to create the mass of the exchanged particle. The process allows the exchanged particle to exist for a very short time in violation of energy conservation. In this case quantum mechanics allows something to happen that would normally be impossible. There are limits, however: the larger the mass of the exchanged particle, the shorter the time the exchanged particle can be in flight. Consequently, higher-mass exchanged particles have shorter "flight times," and they travel shorter distances.

According to Yukawa's theory, the typical distance that the exchanged particles can travel is the "size" of a nucleon. As mentioned above, the distances depend on the mass of the particle. Since the diameter of a nucleon is about 3×10^{-13} cm, the mass of these new particles was estimated to be a few hundred times the electron's mass. The proposed particles would be intermediate in mass between the electron and the nucleons and were called *mesons*.

There was a problem in observing mesons in the laboratory. Electrons, protons, and neutrons are always present within atoms and only need to be kicked out of the atoms to be detected. On the other hand, mesons do not exist permanently within an atom, and they need to be created in order to be seen. The mass (m) of a meson can be created from an amount of energy (E) given by $E = mc^2$. In 1935 no "atom smashers" were capable of delivering enough energy to a nucleus to produce a meson. Cosmic rays, however, are high-energy nuclei that arrive from outer space and bombard the earth's atmosphere. Cosmic-ray nuclei have plenty of energy to produce mesons when they collide with nitrogen or oxygen nuclei in the air.

In 1937 some new particles, known as *mu mesons* or *muons*, were discovered in the cosmic rays. Muons have a mass of 206.8 times the mass of the electron. At first muons were thought to be the particles predicted by Yukawa, but muons do not interact by the strong force, so they are not Yukawa's mesons. Muons are so abundant in the cosmic rays that dozens of them pass (unnoticed) through everyone's body each second!

In 1947 still another type of particle was identified using cosmic rays, called the *pi meson* or *pion* (pronunciation: like py-ahn). Pions have masses of about 270 times the electron mass and do transmit the strong force needed to bind nucleons together within nuclei. Pions fulfill at least part of the role assigned to mesons by Yukawa. Pions have charges of -e, zero, and e, that is, they can have the same (negative) charge as an electron, or no charge, or a charge that is equal in magnitude to that of an electron, but positive.

ANTIPARTICLES, NEUTRINOS, AND PHOTONS

Paul Dirac (1901–1984) developed an important mathematical equation in 1928 that described how electrons interact with electromagnetic radiation. This equation suggested that, besides the usual negatively charged electrons, *positively charged electrons* could also be created. In 1932 D. H. Anderson and R. A. Millikan discovered positive particles, otherwise similar to electrons, in the cosmic rays. These particles were called *positrons*. The positron mass is the same as that of electrons. If a positron is allowed to contact matter it attracts an electron and the two particles *annihilate each other, producing two gamma rays!*

A particle with the opposite charge of another particle of the same mass, but which annihilates the other particle on contact, is called an *antiparticle*. Other antiparticles were sought after the positron's discovery. In 1936 both positive and negative versions of the muon were found. It was expected that antiprotons would be found. They are extremely rare in cosmic-ray interactions in the atmosphere, but a high-energy particle accelerator (commonly called an "atom smasher") was built in order to produce antiprotons. They were detected at the University of California Radiation Laboratory at Berkeley in 1955. Antineutrons were detected in 1956. It is now believed that all particles have antiparticles.

Anti-hydrogen atoms have been formed by bringing together antiprotons and positrons. Presumably, antimatter versions of any kind of object could exist. It has even been suggested that some very distant galaxies might be made of antimatter. There has been no evidence, however, that this is actually the case.

Another interesting and important particle is the *neutrino*. The emission of beta rays by radioactive nuclei was studied in the 1920s. A problem developed because the emitted electrons carried varying amounts of energy, as if they were sharing energy with another undetected particle. In 1931 Wolfgang Pauli (1900–1958) proposed a new particle, the neutrino, as the particle that carried away part of the energy in the emission of beta rays. It was supposed to have no electric charge and would not be affected by strong forces. With these properties, the ghostly neutrino would be nearly undetectable in experiments. In fact, neutrinos interact so little and so capriciously with the natural world that they initially seemed almost mythological!

In spite of the difficulties, the detection of neutrinos was reported by Frederick Reines and Clyde L. Cowan Jr. in 1953, and they produced even stronger evidence in 1956. Neutrinos are so penetrating that they usually pass through the entire earth without interacting. They can also travel from the center of the sun to the earth, where they are observed in underground laboratories in studies of the nuclear reactions which generate the sun's energy. Neutrinos either are massless or have very small masses. It is expected that there are three kinds of neutrinos. As of 1999, results from a huge underground particle detector in Japan and other experiments supported the conclusion that neutrinos do have a tiny mass.

The next important particle is extremely common, although it has some very mysterious properties. It is the *photon*. It is the massless particle that transmits electric and magnetic forces. It is present in all kinds of electromagnetic radiation. Photons transmit energy as radio waves, infrared radiation, light, ultraviolet radiation, X-rays, and gamma rays.

We tend to think of electromagnetic radiation as being wavelike in nature since there is a characteristic wavelength associated with this sort of radiation. It is a strange fact, however, that light and other forms of electromagnetic radiation exhibit both wavelike and particlelike behavior. Which behavior is observed in a particular experiment depends on what type of question the experiment is designed to answer. Experiments that detect the transfer of energy or momentum show the particlelike behavior that arises from individual photons. Experiments that observe the spatial patterns formed by large numbers of photons show wavelike properties, such as interference and diffraction. The particle nature of electromagnetic radiation was discovered in part by Max Planck (1858–1947) in 1900 and in part by Albert Einstein (1879–1955) in 1905.

PARTICLE LIFETIMES AND THE WEAK FORCE

Some of the elementary particles that we have described live forever, or at least an extremely long time. In this category are neutrinos, photons, electrons, and protons. The neutron is interesting in that it can live almost indefinitely in many nuclei, but it decays with a mean lifetime of about nine hundred seconds outside the nucleus. When it decays, it changes into a proton, an electron, and an antineutrino.

The neutron has just a little more mass than the sum of the masses of the proton and the electron, and the antineutrino has no mass. Consequently, there is enough mass-energy for this decay to happen, with the small excess mass-energy being converted into kinetic energy of the three resulting particles. When a neutron enters a nucleus, however, some of its mass-energy is lost when it falls into the range of the strong force that binds it to the nucleus. If the binding is strong enough, the neutron no longer has any excess mass-energy. So the neutron does not decay and it can stay in the nucleus indefinitely.

The pion and the muon are also unstable particles. A charged pion decays into a muon and a neutrino. The mean lifetime of the charged pion is 2.6×10^{-8} seconds. The neutral pion almost always decays into two gamma rays in a mean lifetime of only 8.4×10^{-17}

seconds, radically shorter than the lifetime of the charged pion. The muon has a lifetime of 2.2×10^{-6} seconds. It usually decays into an electron, a neutrino, and an antineutrino.

When two protons collide with each other they usually interact with each other. If we consider how long the protons overlap each other, at nearly the speed of light, the time is only about 10^{-23} seconds. This means that the so-called *strong force* is strong enough to make something happen during this extremely short time. If we look at the lifetime of the neutron, the muon, or the charged pion, the times are many times longer than 10^{-23} seconds. This is because these processes occur by a much weaker nuclear force, *the weak force*.

We noted above that the pion transmits the strong force. The photon transmits the electromagnetic force. What transmits the weak force? Julian Schwinger proposed a theoretical model in 1957 that explains the weak force, while also explaining the electromagnetic force. In this model the weak force occurs by the exchange of two new charged particles, the W+ and the W-. The electromagnetic force is due to the exchange of photons, as in previous successful models.

In 1961 Sheldon Glashow added the Z° to the Schwinger model, thereby removing certain problems. The theory was improved further by Steven Weinberg and Abdus Salam in 1973. At this point the masses of the W+, W-, and the Z° were predicted by this theory, called *the electroweak theory*. The W+ and the W- were predicted to have a mass of 87 ± 2 proton masses and the Z° was predicted to have a mass of 98 ± 2 proton masses.

By 1979 many experiments had given indirect evidence that the electroweak theory was correct and Glashow, Salam, and Weinberg were awarded the Nobel Prize in physics for this theory. The actual detection of the W-, W-, and the Z°, however, occurred in 1983 at CERN (the European Center for Nuclear Research) in Geneva, Switzerland. A team of hundreds of physicists, led by Carlo Rubbia, used colliding beams of protons and antiprotons to generate a few particles of each type. The masses of these particles were consistent with the earlier predictions.

MORE-ELEMENTARY PARTICLES:
THE QUARKS AND GLUONS

Although it seems that we have discussed many kinds of particles, there are still many more that have been discovered. In 1975 physicists at SLAC (the Stanford Linear Accelerator) discovered a kind of big brother to the electron and muon, called the *tau*. Then there is a long list of relatives of the pions and of the nucleons. These were discovered in the 1950s and 1960s. Just when nature seemed to be too complicated, physicists found a simpler reality beneath the elementary particles. As usual, evidence for new particles was found! The new particles gave scientists a more basic understanding of nature, and a simpler set of basic building blocks.

The first step toward the new understanding of basic particles was the classification of the observed particles into groups. In 1961 Yuval Ne'eman and Murray Gell-Mann showed that the mesons and the baryons (heavy particles, including nucleons and their relatives) fall very naturally into patterns of eight particles. In 1962 Ne'eman and Gell-Mann independently concluded that a tenth particle should exist to complete another set in which nine particles had been observed.[3] In 1964 the predicted particle, called the $\Omega-$, or *omega minus*, was discovered at Brookhaven National Laboratory on Long Island and at CERN. The simplicity of the observed patterns led theorists to search for an underlying cause of this simplicity. The simplicity is due to the fact that the mesons and baryons are not true elementary particles but are made up of more basic particles.

About 1964 Murray Gell-Mann and George Zweig independently realized that three types of more fundamental particles could be combined to produce the observed mesons and baryons. These particles were given the fanciful name "quarks." (The "quar" in quark is pronounced as in "quart.") The three types or "flavors" of quarks are called the *up*, *down*, and *strange* quarks, with symbols u, d, and s. In this picture, the mesons are built of combinations of a quark and an antiquark (the antiparticle of a quark). For example, the positive pion is a combination of a u quark and a d antiquark. The nucleons consist of three quarks. The proton has two d quarks and one u quark, while the neutron has one d quark and two u quarks. Just remember Gell-Mann's succinct recipe for making a nucleon: "Take three quarks."

The electric charges of the quarks are unlike other observed particles in that the charges are not whole numbers times e, the electron charge. The u, d, and s quarks have charges of $2/3\ e$, $-1/3\ e$, and $-1/3\ e$, respectively. The quarks are also different from other subatomic particles in that they are not observed as independent particles. They are always confined within the other particles that they form. Scattering high-energy electrons on protons at SLAC gave additional support for such subparticles within the proton. The quark model, in spite of its novel properties, is accepted by particle physicists because much experimental evidence supports it.

Besides the three quark flavors originally proposed, evidence has been obtained for three more kinds of quarks, called *charm*, *bottom*, and *top*, with symbols c, b, and t. Evidence was found for the c in 1974, for the b in 1977, and for the t in 1995. Surprisingly, most theorists now think that no other kinds of quarks will be found!

The quark model suggests the question: "What holds the quarks together when they make up a particle?" (Do you feel as if you've been here before?) A theory, called *quantum chromodynamics*, deals with the forces between quarks by the exchange of particles called *gluons* (pronounced like "glue on," with the accent on "glue"). In the electromagnetic force, photons are exchanged between charged particles. In quantum chromodynamics (denoted by QCD), gluons are exchanged between quarks. In QCD, the quarks carry a new kind of charge, somewhat analogous to electric charge, called *color*. Unlike electrical charge, of which there is only one kind, there are three colors, and there are rules for combining different colors in producing particles.

So, we have gone from atoms to subatomic particles to quarks and gluons. You may be relieved to know that no deeper level of reality has been observed. In my attempt to be brief, the discussion of the discovery of the particles may have been somewhat misleading: only the successes in finding particles were described here. Misleading or mistaken results were ignored. I described only the first convincing detection of each particle, ignoring many other experiments that supported the detections and added to our knowledge of the particles. In reality, there is much more evidence for the existence of the subatomic particles than was mentioned above.

THE FOUR FUNDAMENTAL FORCES IN NATURE

At this point we have introduced the "basic building blocks" that make up the universe, but we mentioned the "cement" only in passing. Now let's briefly discuss what happens to these building blocks. Particles and objects stay at rest or move in straight lines at constant speeds unless they are acted on by forces. Forces produce accelerations of particles or objects in the direction of the forces. The magnitudes of the accelerations are equal to the net forces on the objects divided by the masses of the particles or objects that are being accelerated. If object A exerts a force on object B, then object B exerts a force of equal magnitude but in the opposite direction on object A.

The rules given above are Newton's laws of motion. They are accurate except for cases that are far from our everyday experience. Exceptions occur for sub-microscopic particles, or for particles traveling near the speed of light or in the vicinity of enormous gravitational fields. Under such circumstances the rules are modified, but physics can still treat these situations.

What are the forces that exist in the world? Only four kinds of forces are known. These forces all involve interactions between elementary particles, but they affect everything that happens in the large-scale world in which we live. The four forces are: the strong force, the weak force, the electromagnetic force, and gravity. Each will be discussed briefly.

The strong force is a very powerful, but very short-range force, with a range of about 10^{-13} cm. It binds quarks together to form nucleons and pions. It is also responsible for holding the nucleons together within the nucleus. Since the chemical properties of atoms depend upon the total number of protons held together in the nucleus, the strong force is indirectly necessary for the development of the complicated chemical phenomena that make life possible.

The strong force plays another all-important role in our lives. Almost all of our energy arises from nuclear interactions that occur within the Sun. This solar energy produces the chemical energy in our food and our fuels. It powers the evaporation of water that generates our hydroelectric power as well. About the only major source of energy that is not ultimately derived from solar nuclear energy is nuclear energy produced artificially on Earth! Apart from its applications in nuclear energy and nuclear warfare, the strong force does not play a large role in producing change in our world. So it is rarely involved in the processes that are reported in the evening news. If it does make the news it could be really bad news!

The electromagnetic force produces the attraction or repulsion between electrically charged objects. It also produces magnetism and all the forms of electromagnetic radiation. This interaction is involved in most events of our lives. The matter in our ordinary environment is in the form of atoms or molecules. Atoms hold together by the attraction of the opposite charges of the nuclei and the orbiting electrons. Molecules are atoms that are held together by electrical forces.

When you touch a table and it feels hard, the electrons of atoms within the surface of your skin strongly repel the electrons in the atoms on the surface of the table, keeping your finger from penetrating the surface of the table. When you see a distant object, you detect light reflected or emitted by the distant object. This light is electromagnetic radiation traveling from the distant object to your eye. Electrical forces cause the chemical behavior of atoms. Consequently, electrical forces participate in all the chemical processes in your body.

The weak force produces the beta-ray emission by nuclei and the decay of some of the unstable elementary particles. Like the strong force, it is involved in the processes of energy production by the Sun and other stars. It also affects the formation of the elements within stars. Like the strong force, it is usually not involved in the changes occurring in our ordinary lives.

Gravity is the most obvious of the forces. It was the first fundamental force to be described successfully by physics. It is quite weak, but it has an unlimited range. Forces between electrical charges are much stronger and also of unlimited range, but there is very little total electric charge on most large objects. As a result, the long-range effects of the electrical force are usually small. On the other hand, gravity attracts other matter indiscriminately. There is no way to hide or neutralize the effect of gravity on a massive object. Consequently, gravity is usually the most important force between astronomical objects.

Gravity holds us, our water, our air, our vehicles, and our houses on the earth's surface. It holds the planets near the Sun. It collects the sparse matter in space into galaxies, where densities can become large enough for matter to concentrate even more to form stars and planets. Unlike the other forces, no particle has been found which transmits gravity. Such a particle, called a *graviton*, may exist, but it has not been detected.

HOW PHYSICS UNDERSTANDS THE PHYSICAL WORLD

The four forces and the subatomic particles, together with Newton's laws, quantum mechanics, and relativity, form the basis for understanding all physical phenomena. Heat, light, electricity, magnetism, atoms, nuclei, gases, liquids, solids, and the motions of objects are all at least partly understood in these terms. What is not understood in these areas is thought to be *understandable in principle* in terms of these forces, particles, and theories. Future knowledge will reach deeper levels, but the present state of knowledge is very impressive.

All of the phenomena of chemistry are also understandable in principle in terms of these fundamental properties of matter. Examples include the relative amounts of materials used in different interactions, the existence of elements, the periodic table, and the forces between atoms that cause chemical compounds to form.

Scientists have found that all of the processes of biology are consistent with the fundamental properties of matter. For example, we do not need to propose a fifth fundamental force in order to explain how muscles generate force. Muscles generate force by the usual laws of chemistry and physics, by a process that will be described in chapter 7.

By saying that biological processes proceed according to the laws of physics and chemistry, I am not saying that biology is reduced to just physics and chemistry. There are a number of reasons for this. One reason is that much of biology deals with facts that are related to prehistoric accidents that are not, at least in practice, predictable on the basis of physical theory. For example, large asteroids that struck the earth in the past probably caused the extinction of large numbers of species. The details of how big these asteroids happened to be and when and where they struck Earth played a large role in determining which forms of life now exist on Earth. It seems likely that science will never be able to predict these details. Thus, a large part of biology involves observations of nature to learn about these unpredictable facts.

Another reason why biology is not reduced to physics is that biology deals with subject matter that is more restricted in scope than physics and chemistry. That is, it usually deals with living matter on our particular planet, not all matter. With these restrictions and the special characteristics of living matter, many properties and rules exist which apply to living matter on Earth, but not to all matter. These properties and rules emerged from matter because special circumstances apply in Earth's particular environment. The special circumstances on Earth include the abundance of water, a temperature range that makes liquid water abundant, the presence of a thick atmosphere, etc. Under these special circumstances, life-forms developed, and the theory of evolution seems to describe much of this devlopment. Such emergent properties and emergent rules, such as the survival of the fittest, result from the fact that biology is a higher level, or more specific, science than physics or chemistry.

Although biological processes can be reduced to physics and chemistry in principle, in actuality this is often not practical. Even when this reduction is practical, the reduction to physics and chemistry is normally not a sensible way to teach or communicate biological knowledge, because the procedure of reducing biological facts to physics or chemistry is complicated and tedious. Similar arguments explain why other sciences are not reduced to physics and chemistry.

Now let's jump from biology to the much more restricted subject of processes within the human brain. This is a shift up at least three levels from biology. Animals are only part of biology, nervous systems are only a part of animals, and we are considering only human brains, not nervous systems in general. With this great increase of level, it is not surprising that major differences exist in the rules and properties of human brains, compared to life in general.

The human brain has something like 10^{11} neurons, similar to the number of stars in a galaxy such as our own Milky Way. Each neuron is connected with thousands of other neurons within the brain. This enormous network of neurons allows us to display very complicated behavior. The brain makes the atom look enormously simple by comparison. The brain is a learning machine that has acquired knowledge from all the experiences of a person's life, and it is perhaps the most complicated thing in the known universe. It is not surprising that the brain has wonderful abilities!

Everything that happens within the brain still involves, at the lowest level, the interactions of systems of elementary particles that respond to the four fundamental forces. At the lowest level, nothing new has been introduced. The functioning of the brain could be described, in principle, in terms of the orderly forces operating between elementary particles. Looked at from this level, the brain is an impersonal material object.

The person emerges when we consider higher levels, and is related to the detailed organization of material in this particular object, the brain of an individual human being. Should we be surprised that a brain consisting of 10^{11} active components and perhaps 10^{14} or 10^{15} connections, with millions of years of testing and improvement, should be able to compose music or write poetry? Such a wonderfully organized and complicated device ought to be capable of directing all of a person's activities, and I believe that is what it does.

NOTES

1. Stephen F. Mason, *A History of the Sciences* (New York: Collier Books, 1962), p. 33.

2. H. T. Pledge, *Science Since 1500* (London: Her Majesty's Stationery Office, 1966), p. 217.

3. Frank Close, *The Cosmic Onion: Quarks and the Nature of the Universe* (New York: American Institute of Physics, 1983), p. 66.

6

OUR COSMIC GENESIS I
The Origin of Matter and the Earth

I N THE PREVIOUS CHAPTER I described the basic particles and laws by which the universe operates. The *history* of the universe, however, is just as important to us as the properties and interactions of the fundamental particles. In order to present a worldview that allows the basic arguments of this book to be made, it helps to give an alternative to the well-known Old Testament account of the origin of the world. In this chapter I will summarize scientific knowledge about the fascinating history of matter, the Solar System, and Earth. The raw materials for these histories are not historical documents, of course. They are observational facts along with the scientific theories that explain and organize these facts.

Our universe consists of all the space, matter, and energy that we can possibly observe or study. We know there was a time when the organized systems of elementary particles, such as stars and human beings, did not exist. Scientific theory can follow the universe's history back to an extremely hot "soup" (called a *plasma*) of fundamental particles before even atoms and nuclei existed. Scientists are even speculating on how the fundamental particles originated. Let's consider the trail of evidence and model-making which scientists followed when they developed their theory of the origin of the universe.

GALAXIES

The discovery of *galaxies* was an important step in uncovering the universe's history. Our own galaxy was the only galaxy known until about 1924. It is an enormous disk-shaped group of about 10^{11} (100 billion) stars. We are located within this disk, but far out from the center. When we look long distances through this disk in a dark night sky, we see the light from billions of stars. With the naked eye we can't distinguish most of the individual stars, but we see a long broad trail of light: the Milky Way. From our viewpoint within the disk, the Milky Way traces a complete great circle in the sky. This special pattern was appreciated by the great German philosopher Immanuel Kant (1724–1804). He wrote the following in 1755, long before the nature of the Milky Way was known with certainty.

> Whoever turns his eye to the starry heavens on a clear night, will perceive that streak or band of light which on account of the multitude of stars that are accumulated there more than elsewhere, and by their getting perceptibly lost in the great distance, presents a uniform light which has been designated the *Milky Way*. It is astonishing that the observers of the heavens have not long since been moved by the character of this perceptibly distinctive zone in the heavens. . . . It is seen to occupy the direction of a great circle, and to pass in uninterrupted connection around the whole heavens: two conditions which imply such a precise destination and present marks so perceptibly different from the indefiniteness of chance, that attentive astronomers ought to have been thereby led, as a matter of course, to seek carefully for the explanation of such a phenomenon.

Besides the Milky Way, the Andromeda Galaxy and the Large and the Small Magellanic Clouds were visible to people before the telescope was invented. (The Magellanic Clouds are two small galaxies named after Ferdinand Magellan, who observed them in 1519 during his famous trip.) Many other galaxies were visible after the invention of the telescope. Kant, in a perceptive and poetic identification, called them "island universes." The problem for scientists, if not for philosophers, was that these fuzzy patches of light, called *nebulae*, had not been shown by science to be other galaxies. It was not known that these nebulae are great distances from us. The stars within these galaxies were not distinguishable with early telescopes because the distances to these galaxies are so great. The problem is complicated because there are different kinds of nebulae, some are located within our galaxy, and others are distant galaxies.

In the early 1920s the American astronomer Edwin Hubble (1889–1953) detected individual stars within the Andromeda nebula and found a way to estimate their distance. These distances were so enormous that Hubble correctly concluded that this nebula is really a separate galaxy, the Andromeda Galaxy. Soon Hubble and others identified very large numbers of galaxies. The size of the known universe increased drastically with the discovery of these distant galaxies. Distances to "nearby" galaxies are on the order of millions of light-years. A *light-year* is the distance light travels in one year. The speed of light is about 300,000 kilometers per second (186,000 miles per second). To help you imagine the scale of a light-year, the distance to the moon is 1.3 *light-seconds*.

THE BIG BANG

The next step in "reading" the history of the universe came from observing the starlight emitted by distant galaxies. The surfaces of stars contain gas at very low pressure. Within this gas, atoms of each chemical element produce or absorb light at certain specific wavelengths. Astronomers use prisms to separate starlight into its spectrum of colors. The distribution of the intensity of light of different hues is called a *spectrum* (plural: spectra). Within the spectrum certain wavelengths corresponding to certain elements are outstandingly bright or dark, telling astronomers which elements are present in the stars. Thus, astronomers can learn which kinds of atoms are present in stars that are thousands of light-years away!

The spectra can teach us something else about the galaxies. If a galaxy is moving toward us, the characteristic wavelengths corresponding to specific chemical elements are shifted toward the blue, or shorter wavelength end of the spectrum. A galaxy that is moving away from us has the wavelengths shifted toward the red, or longer wavelength end of the spectrum. In 1929 Hubble showed that most galaxies have their spectra shifted toward the red. For random motions, the number of galaxies that are red-shifted would be roughly equal to the number that are blue-shifted. Hubble found that most galaxies are moving away from us and that the *speed of this motion is proportional to the distance of the galaxy.*

The observations suggested that the entire universe is expanding, so that the distance between pairs of widely separated galaxies is increasing at a rate that is proportional to the separation between the galaxies. One analogy is with a huge loaf of raisin bread. As the dough is cooked the bread expands and the raisins get farther apart. A pair of raisins that are close to each other separate slowly while a pair of widely separated raisins separate more rapidly. Another analogy is with an explosion of an object. After the explosion, the fragments of the object separate like the raisins or galaxies described above. The explosion scenario suggests the name of this picture: the *big bang* model. The explosion analogy also suggests, correctly, that matter was very hot initially and that it has cooled as the universe expanded.

A basic parameter of the big bang model *is the speed at which distant objects move away from us divided by the distance of these objects.* This ratio is known as *the Hubble Constant.* The speed is given in kilometers per second. The distance is given in *megaparsecs.* One megaparsec equals 3.26 million light-years, an enormously long distance. The value of the Hubble Constant is not known accurately: values from fifty to ninety kilometers per second per megaparsec have been reported in the last twenty years. A recent (1999) value obtained with the Hubble Space Telescope is seventy-one (with an uncertainty of seven) kilometers per second per megaparsec.

If we make the crude assumption that the separation speeds of matter that is now in galaxies has been constant since the original big bang, we can estimate the approximate time since the matter was at the same place; that is, when the big bang occurred. Using seventy for the Hubble Constant gives 1.4×10^{10} years, or fourteen billion years. Using fifty gives twenty billion years.

There are other ways to estimate the age of the universe. An estimate of the age of uranium, based on the abundances of surviving isotopes, gives about ten billion years. The uranium was produced at some time *after* the big bang, so the universe must be at least ten billion years old. Experts on the evolution of stars conclude that the oldest stars in our galaxy are about fifteen billion years old. The age of the universe must exceed the age of the oldest stars, so this result suggests an age of at least fifteen billion years, toward the upper end of the range suggested by the Hubble Constant. In any case, the different methods give results that are reasonably consistent with each other.

For times earlier than an extremely tiny fraction of a second after the big bang, the history of the universe is uncertain. If we push the big bang model back to such early times, it predicts enormously high temperatures and nearly infinite densities of matter and energy. I will mention some speculative ideas about these earliest times later. For now, let's consider the time interval when atomic nuclei and atoms were produced. This period seems to be well understood.

About *one second* after the big bang, the temperature was about 10^{10} K; that is, ten billion degrees Kelvin. (The Kelvin scale has degrees of the same size as the centigrade scale, but the Kelvin scale is shifted so the lowest possible temperature is zero.) The universe was filled with plasma at this very high temperature. Protons, neutrons, electrons, positrons, photons, and neutrinos were abundant. No nuclei existed since the photons (in the energy range where they are called gamma rays) were so energetic that they were able to break up any nuclei by colliding with the nuclei. Equal numbers of neutrons and protons were present.

By *a few minutes* after the big bang, the temperature had decreased to less than one billion degrees. Most of the neutrons had converted to protons by interactions and decays, so there were about seven times as many protons as neutrons. At this time *the temperature was low enough that nuclei could form* and they would usually not be destroyed by collisions. In the first step, collisions of protons and neutrons in which the particles stuck together formed deuterium nuclei. Then collisions involving deuterium nuclei produced nuclei with three nucleons. Other collisions generated helium nuclei with four nucleons (two protons and two neutrons). Here the process nearly stopped since there is no stable nucleus with five nucleons. Another barrier exists at eight nucleons: there is no stable nucleus with eight nucleons.

About 25 percent of the nucleons were bound together in helium nuclei at the end of this nucleus-building process. Most of the remaining nucleons were protons (hydrogen nuclei). The relative abundances of some nuclei remain about the same as they were after the first few minutes of the big bang. The big bang model predicts the fractions of helium with three and four nucleons that were produced, along with the trace amounts of deuterium and lithium. *In one of the triumphs of the big bang theory, all of these predicted abundances are in quite good agreement with the observed universal abundances. Thus, science has made sense of what happened in the first few minutes after the formation of the universe!*

Except for hydrogen, most atomic nuclei of elements that make up living matter did not exist at this stage. These nuclei were produced a billion or more years later in stars. In the first years after the big bang, even the nuclei that *had* been produced did not form atoms, because the temperature was too high for atoms to hold together. Electrons were available to combine with nuclei to form atoms, but the electrons in atoms were immediately separated from nuclei by collisions with energetic photons.

About 300,000 years after the big bang, the temperature had dropped to about 3000 K and electrons were able to combine with nuclei to form atoms since the photons were less energetic at this temperature. This change from a universe filled with plasma to a universe filled with atoms had a major effect on the photons. The universe had been opaque since the photons could not travel long distances without being scattered by collisions with free electrons. After electrons were bound into atoms, however, the photons did not scatter off the atoms and *the universe became transparent*. In fact, the universe became so transparent that this radiation from shortly after the big bang is still detectable after traveling for fifteen to twenty billion years!

The existence of this fossil radiation from the big bang is one of the strongest pieces of evidence supporting the big bang model. Even before the radiation was discovered, it was known that the wavelengths of the radiation should be increased by a factor of about a thousand relative to the original wavelengths. This increase is due to the enormous expansion of the universe that has occurred since the radiation was emitted. The radiation is now in the microwave region of the spectrum. When the radiation was emitted, it had a typical spectrum of a very hot gas at 3000 K. When it is received, at present, it looks as though it was emitted by an extremely cold body with a temperature of 3 K. (This is just three degrees above absolute zero, the temperature at which all thermal motion ceases.) The temperature was reduced by the same factor of 1000 by which the universe had expanded.

The *universal 3K radiation*, often called the *cosmic background radiation*, was discovered by accident by Arno Penzias and Robert Wilson in 1964. They were using a very sensitive microwave detector at Bell Telephone Laboratories. The detector was designed to receive signals from a satellite. The radiation appeared as a low intensity "noise" with equal intensity from every direction in the sky. Penzias and Wilson removed all conceivable sources of this noise, including pigeon droppings on the antenna, and the noise still remained. They learned later that this radiation was expected as a product of the big bang. Because the radiation had the expected temperature and was present in all directions, it was rapidly accepted as *a fossil remnant of the early universe*. The big bang theory had scored again with another brilliant prediction!

A few years ago, a sensitive NASA-funded satellite experiment called the Cosmic Background Explorer, or COBE, measured the intensity of the microwave radiation at fifty wavelengths. These wavelengths covered essentially the whole spectrum present in 3 K radiation. The observed intensities agree, to an accuracy of 0.03 percent, with the shape expected from a perfectly absorbing body, or *blackbody*. In the big bang model, the blackbody spectrum is just what was expected to result from the furnacelike conditions that existed when the radiation was emitted. The COBE experiment measured the precise temperature of the "3 K" radiation: it is 2.726 K.

The most famous result of COBE is the discovery in 1992 that the radiation is not quite isotropic. That is, it has been observed that the radiation does not have precisely the same temperature in all directions. Before those deviations could be found, however, another effect needed to be removed from the data. The radiation serves as a "cosmic speed detector." Any motion of the earth relative to the whole universe causes a blue shift of the 3 K photons arriving from the direction of the earth's motion, with a corresponding red shift of photons arriving from the opposite direction. COBE observes such an effect. It is due to a *620 kilometer per second* motion of our galaxy toward the constellations Hydra and Centaurus. This motion is thought to be caused by the gravitational attraction of an especially high concentration of galaxies in that direction.

After the effects of the earth's motion were removed from the COBE data, there were still many regions of the sky with higher or lower temperatures than the average sky temperature. These regions are patches about ten degrees or larger in angular diameter. The temperature deviations are tiny: about fifteen millionths of a degree. These deviations are attributed to slight variations in the matter density in the early universe.[1]

Some such deviations were predicted since 1967, decades before they were found. They were expected since matter is not distributed uniformly in space. Even at size scales beyond the scale of galaxies, there are *clusters of galaxies* and even *superclusters* containing clusters of galaxies. Enormous regions that are almost empty of galaxies have also been observed. There had to be matter-density fluctuations in the early universe that grew with time to produce the present density variations. On the other hand, the 3 K radiation is still remarkably isotropic since the temperature is uniform in all directions to a part in 10,000. It was a great feat to observe the fluctuations at all!

INFLATION AND THE ORIGIN OF MATTER

Let's discuss some speculative issues concerning the origin of the universe. Science can't deal confidently with what happened at an extremely tiny fraction of a second after the big bang when the universe possessed enormous temperatures and densities. Scientists aren't sure how matter behaves when exposed to these extreme conditions. Although experimenters have not tested matter under such conditions, theorists have some promising ideas about what might occur under these circumstances.

About 1979, the American physicist Alan Guth found that a somewhat speculative fundamental theory that can give a combined description of the strong, weak, and electromagnetic forces has major implications for the extremely early development of the universe. Guth developed a theory, called *inflation*, that claims that between 10^{-35} and 10^{-33} seconds after the origin of the universe certain processes caused a strong repulsive force to arise between particles, a kind of "negative gravity." This force caused the universe to explode in a runaway and exponential fashion, doubling in size each 10^{-35} second. At the end of this process the expansion rate reduced to the rate given by the big bang model.

Inflation solves some problems. One problem is to explain how all of the universe started expanding at the same time, since the start of expansion had to happen at such early times that no causes or signals could have traveled between separated parts of the universe. In the inflation picture the expansion was so rapid at first that the universe could have been "synchronized" since all of the observable universe was in possible communication with itself within a very small volume just before the onset of inflation.

Inflation also seems to make a "free lunch" universe possible. Matter was made from energy. What was the source of all the energy that was needed to produce the matter in the universe? The answer seems too wonderful to be true. The energy that went into matter is balanced against an equal amount of negative gravitational energy that is produced by the existence of this matter. *No net energy is needed to "create" the mass.* It appears that matter could come into existence from nothing. *The whole universe may be a free lunch!*

The universe may have started as a random fluctuation in empty space. One originator of this idea was an American physicist, Ed Tryon.[2] In 1973 he published a paper in *Nature* entitled "Is the Universe a Vacuum Fluctuation?"[3] He realized that quantum mechanics, through the Heisenberg uncertainty principle, allows particle-antiparticle pairs to be produced randomly, spontaneously, and momentarily from the vacuum. With the addition of energy, these pairs could be brought into long-term existence. Tryon also realized that the produced gravitational energy could compensate for the particle-production energy so that matter formation required no input of energy. *Together, these two ideas seem to allow the universe to originate by itself.*

Since Tryon's work other theorists have proposed models for the spontaneous origination of the universe. The most famous model is the *no boundary proposal* by the American scientist Jim Hartle and the famous British scientist Stephen Hawking. In general, the philosophy of these models is to treat the origin of the universe as a result of the laws of physics. Previously, many scientists viewed the origin as an act of creation or as a process that was not understood. The laws of physics were thought by many to apply only *after* the origin. In these new models, the universe comes into existence in a way that is consistent with the laws of physics.

Two other important problems connected with the origin of matter may share a single solution. The problems involve the relative numbers of different kinds of particles in the universe. One problem (the *photon problem*) is that there are about a billion times as many photons as nucleons. Theorists would expect the ratio of photons to nucleons to be much closer to one-to-one. The other problem (the *antimatter problem*) is how to explain the excess of matter over antimatter in our universe. If we imagine that matter originated by a process in which pairs of particles and antiparticles were produced together, we would have equal numbers of particles and antiparticles at the time of production. The Solar System, however, has almost no antiparticles and vast numbers of particles. *Where have all the antiparticles gone?*

A conceivable answer to the antimatter problem is that matter and antimatter were separated by some process. If antimatter is abundant elsewhere within our galaxy we should detect some antinuclei among the cosmic rays. These antinuclei have not been detected, except for a few antiprotons, which were probably produced recently in high-energy collisions.[4] If some galaxies are made of antimatter, collisions between antimatter and matter galaxies would generate enormous amounts of gamma radiation from annihilation processes. Very many pairs of colliding galaxies are observed, but none display evidence of annihilation of matter with antimatter. At present there is no convincing evidence that any other galaxies are made primarily of antimatter.

It seems more likely that the universe now contains very little antimatter. If there were equal numbers of particles and antiparticles in the very early universe, something must have happened which selectively removed the antiparticles. This is surprising since it seems more natural to assume that all physical processes are the same for matter and antimatter.

In 1964 American physicists James Cronin and Val Fitch showed that certain decay rates of neutral K mesons are slightly different for particles and antiparticles. Thus, the symmetry between particles and antiparticles is violated, to use the language of particle theorists. If nature does not preserve the particle/antiparticle symmetry in the case of neutral meson decays, nature may also violate that symmetry in producing matter and antimatter in the early universe. It seems likely that as the very early universe cooled, a rare process generated or preserved slightly more matter than antimatter. In the end *almost all of the matter and all of the antimatter annihilated each other*, producing abundant photons and leaving a relatively small amount of matter. This explanation solves both the photon problem and the antimatter problem.

THE STARS:
FACTORIES FOR LIFE'S RAW MATERIALS

The Sun, our nearby star, is the ultimate energy source for almost all life on Earth. But life depends on stars in another way that is less obvious. Except for hydrogen, all of the chemical elements necessary for life were produced by stars. *Life on earth formed from materials produced in stars that existed billions of years ago, before the formation of the Solar System.* Near the ends of the lives of these stars, much of their material was ejected into space. This *interstellar gas and dust* was the source of the material that formed the Sun and the planets, including Earth. *Our bodies are made of this gas and stardust!*

Before the stars existed, most matter was hydrogen and helium produced by the big bang. Galaxies formed as matter fell toward regions with higher than average density. Stars formed within galaxies as clouds of gas collapsed by gravity. As this gas collapsed under its own weight, it heated up enormously. When the temperature reached millions of degrees *nuclear fusion processes* started occurring near the centers or *cores* of the stars.

During most of their lives, stars produce energy by fusion reactions in which hydrogen nuclei (protons) combine, by a number of steps, to form helium nuclei. This process releases very much energy. Part of this energy escapes in the form of neutrinos. Another part is converted to heat that eventually reaches the surface of the star, where it is emitted as starlight. During this *hydrogen-burning process* the star produces energy at a fairly constant rate. Our sun has done this for 4.56 billion years. The Sun is expected to continue operating this way until it is about ten billion years old, when all of the hydrogen in the core will have been changed into helium. When the Sun or any other star reaches this point a crisis occurs. *The "nuclear fire" goes out* because of the lack of nuclear fuel that will "burn" at the temperature of the center of the star.

As a result of the lack of hydrogen in the star's core, some readjustments take place so that the core collapses and becomes more dense. This collapse heats up the core until the temperature becomes so high (about eighty or ninety million K) that *the helium nuclei start combining to form larger nuclei*. Nuclear reactions fuse three helium nuclei into twelve-nucleon carbon nuclei. These processes operate quickly and generate energy so rapidly that the outer part of the star increases greatly in size. When this happens to the Sun, it will enlarge so much that it will swallow the inner planets, including Earth! But this shouldn't happen until billions of years from now.

After carbon is produced in the *helium-burning process*, it can combine with helium to make oxygen. I will not describe the remaining processes any further except to say that stars that are much more massive than the Sun can produce all the naturally occurring nuclei by known processes. *Energy released* as nucleons are added to nuclei in the nuclear-fusion reactions that I have been describing.

The process of adding nucleons to nuclei does not go on indefinitely. This energy-producing process continues for larger and larger nuclei until iron nuclei, with about fifty-six nucleons, are made. The nucleons are bound more tightly inside iron nuclei than in any other nucleus. Nuclei that are heavier than iron nuclei can be made, and are made, but *pulling nucleons from one iron nucleus to add them to another iron nucleus to form a nucleus that is heavier than iron is a process that uses up energy*. So eventually a very massive star reaches a point where no more energy is released by nuclear fusion. The continual liberation of energy is required for a star to remain stable, so *the end of energy production by fusion results in a dramatic crisis for the star*.

When the core of the star has no energy source, its temperature and pressure fall. The decrease of pressure allows the outer layers of the star to collapse and fall toward the core by the force of gravity. These outer layers accelerate inward and then collide with the very rigid core and bounce off the core. The bouncing matter sends a shock wave back up through the star. *This shock wave blows off the outer parts of the star*. This rapidly expanding explosion of very hot gas is extremely bright and the star shines extremely brightly for a few months. This process is called a *super-nova*. It is one of the ways that chemical elements produced in a star escape into space. There is another way in which a supernova can occur, but the effect is the same: chemical elements produced in the star are thrown out of the star to serve as the material for the next generation of stars as well as planets.

FORMATION OF THE SOLAR SYSTEM

Newton's laws of motion and of gravitational attraction explained the *motion* of the bodies in the Solar System. Properties of the Solar System other than motion were believed by many to be unchanged since Creation. Some people felt that God needed to correct the motion of the Solar System from time to time, in analogy with a person who occasionally corrects the time displayed by a clock. As often happens in the unfolding of our knowledge of nature, *facts just outside the current range of human understanding were assumed to require a supernatural explanation.*

Some progress toward a more complete scientific understanding of the Solar System was made by the French mathematicians and physicists Pierre Simon Laplace (1749–1827) and Joseph Louis Lagrange (1736–1813). They showed that certain irregularities in the motion of the Solar System are self-correcting. Consequently, the motion of the Solar System is quite stable and does not require supernatural intervention in order to maintain its orderly operation.

Laplace also made a major advance toward an evolutionary model of the development of the Solar System. He noted that all of the planets known at his time orbit the Sun in roughly a single plane. All planets orbit in the same direction and the Sun and most of the planets have spins in nearly the same directions. Laplace proposed that the Solar System formed from a rotating mass of gas. The orderly motions of the planets are a result of conservation of angular momentum in Laplace's picture. Part of Laplace's model survives in the current ideas of the formation of the Solar System.

The modern picture of the development of the Solar System is based on: (1) observations of star formation in our galaxy, (2) the chemical compositions and other properties of the Sun, the Moon, the planets, and meteorites, and (3) computer simulations of Solar System evolution. Studies of the amount of decay of radioactive nuclei in meteorites show that the formation of the Solar System occurred 4.56 billion years ago. The ratios of amounts of certain lead isotopes found on Earth support the conclusion that the earth is also 4.5 or 4.6 billion years old.[5] Recall that the universe formed about fifteen to twenty billion years ago. *Thus, the Sun and the earth were not in existence during much of the history of our galaxy and the universe.*

The Solar System formed near the plane of the galaxy about 30,000 light-years out from the center of the galaxy. On the scale of billions of years, the formation of the Sun and the planets was relatively rapid. Most of the process was finished within about a hundred million years or less. The Sun formed from material that was enriched in elements produced by previous generations of stars. The original material was mainly hydrogen and helium gas, with significant smaller amounts of carbon, nitrogen, oxygen, neon, magnesium, silicon, sulfur, argon, and iron. Still smaller amounts of all other stable elements were present. There were also radioactive isotopes from a supernova that occurred some millions of years before the formation of the Solar System.[6] Whether the supernova played a role in triggering the formation process is not known.

There have been many theories of how the Solar System formed. This is a rapidly changing field and the picture I present may be modified in the future. The Solar System formed from a rotating cloud of gas and dust. Most of the matter in this cloud fell, under the influence of gravity, into the center of this cloud, where the Sun was formed. Some of the dust and gas, carrying angular momentum from the cloud's original rotation, remained in orbit about the rapidly forming sun. The dust gradually collapsed from a more or less spherical shape into a thin disk.

The Sun heated up as material fell into it. Eventually the Sun's center got hot enough to start nuclear fusion. The Sun, like other newly formed stars, may have emitted a strong wind which blew much of the gas away from the inner parts of the Solar System. Solar heating and solar winds may have removed easily evaporated materials from the inner Solar System as well. Some such process may have produced the large differences in chemical composition that exist between the inner planets (Mercury, Venus, Earth, and Mars) and the outer planets (Jupiter, Saturn, Uranus, Neptune, and Pluto). The inner planets are made up mostly of rock and metals such as iron and nickel. The outer planets, except Pluto, have enormous amounts of hydrogen and helium and may have cores of rock and ice. Pluto has almost no atmosphere, and probably consists of rock and ice.

Collisions of dust grains within the Solar System disk produced larger objects as the grains stuck together. These objects were involved in more collisions in which their size increased from dust grains to pebbles to boulders, and so on. Eventually very large solid objects built up within the disk. These objects, called *planetesimals*, became large enough that gravity became important in bringing objects together. The planetesimals, ranging in size from as large as the planet Mars[7] to only a few meters (or yards, if you prefer) in diameter, collided with each other to form planets. The planets and moons that have not been altered by weathering show craters produced during late stages of the bombardment by planetesimals. The last major crater formation on the Moon occurred about 3.85 billion years ago.[8] The planetesimals are now gone from the region of the inner planets, except for a few "party crashers" that arrive from farther out in the Solar System.

The origin of the Moon is interesting and "impacts" on the early history of Earth as well. It is believed that a Mars-sized planetesimal suffered a grazing collision with Earth. Part of the outer, less dense part of the impacting object went on to fall into orbit around Earth, forming the Moon. The remainder was incorporated into Earth. The formation of the moons of the giant outer planets, such as Jupiter, may have been much different. Jupiter has sixteen moons and resembles a miniature solar system. Jupiter and its moons may have formed, like a miniature solar system, from an extended cloud of gas and dust, forming a large central body surrounded by a rotating cloud, and so on.

Nine planets, with a total of sixty some moons, remained at the end of the Solar System's formation. In terms of mass and brightness, the Solar System is dominated by the Sun. In terms of interest for the development of life, the Solar System is dominated by Earth. Aspects of Earth's history related to the development of *life* are emphasized in the next section.

EARLY EARTH

Earth was probably very hot at the time of its formation and it may have been melted, even at the surface. One heat source was the kinetic energy of planetesimals that collided with Earth, including the event that produced the Moon. Another source of heat was the radioactivity of certain isotopes from the supernova event that preceded the formation of the Solar System. Two major kinds of material were separated by gravity in the developing Earth. Relatively heavy metals, mostly iron and nickel, settled to the center of Earth, forming the *core* of Earth. Somewhat lighter, rocky materials formed the *mantle* outside the core. This separation of materials liberated a great deal of energy that also contributed to heating Earth.

On top of Earth's mantle, a solid *crust* of rock formed as Earth cooled. From the abundance of radioactive isotopes, the ages of the oldest rocks have been determined. Some zircons from Western Australia formed 4.2 billion years ago, but they are embedded in younger rock that is "only" 3.5 billion years old. The oldest rock formation known at present is found in northwest Canada and is 3.96 billion years old. Ages of about 3.8 billion years occur in a number of widespread locations. It is possible that most of any crust that was formed very early was destroyed by the bombardment by planetesimals that ended about 3.8 billion years ago.

Not all of Earth's crust formed at once. After the episode of crust formation 3.8 billion years ago, a large amount of continental surface was produced between 2.8 and 2.5 billion years ago. The first really large continents had appeared by the end of this interval.[9] There also was strong growth of continents between 2.1 and 1.8 billion years ago and between 950 and 650 million years ago.

Liquid water seems to have been present as early as 3.8 billion years ago since some crystals incorporated into rocks of this age have rounded edges, apparently due to being transported by water. Planetesimals and comets falling into Earth are possible sources of the water. Earth has remained a favorable environment for life during at least the last 3.5 billion years through the continuous presence of liquid water. In contrast, our nearest planetary neighbors, Venus and Mars, have temperatures far above boiling and well below freezing, respectively.

Earth did not collect its atmosphere directly from the original solar nebula. If it had, it would, like the Sun, have very large amounts of the so-called *noble gases* within its atmosphere. (The noble gases are chemically inert, that is, *they do nothing* when exposed to other chemicals. They are called noble because the property of doing nothing supposedly made them resemble the European nobility.) Rather than forming from gas in the original solar nebula, the atmosphere may have come from gas that was trapped within the solid planetesimals that made up Earth. The gas may have been released to the atmosphere from volcanoes or by other means. In-falling asteroids and comets from the outer Solar System may also have contributed to the atmosphere by carrying in water, ammonia, methane, and carbon dioxide in frozen form, along with rocks and metals.

Earth's atmosphere probably formed very early. It formed more than 4 billion years ago, based on the ratios of certain isotopes of argon and xenon.[10] Initially it probably contained ammonia, methane, water vapor, and hydrogen. Fairly early, however, it changed to nitrogen, carbon dioxide, and water vapor.[11]

A major difference between the early atmosphere and the present one was the lack of free oxygen. Free oxygen is oxygen that is not bound into chemical compounds. The lack of free oxygen is not unusual: the atmospheres of the other planets do not have free oxygen either. Very old sedimentary rocks contain iron in a chemical form which would have been different if there had been oxygen in the sea water in which the rocks formed. This implies there was no oxygen in the atmosphere, either. A change occurred about 2.0 billion years ago when rocks colored "rusty red" by oxidized iron began to appear.[12] This is evidence that oxygen built up in the atmosphere at that time. The source of this oxygen was *photosynthesis*, the process by which green plants take in water and carbon dioxide and produce oxygen and carbohydrates.

The relationships between oxygen and life are interesting. Many of the compounds necessary for the *origin* of life would not have been made in the presence of oxygen. Later, oxygen was *necessary* for animal life. Oxygen was also necessary for life on land since it produces *ozone* in the upper atmosphere which shields life from deadly ultraviolet rays from the Sun. Thus, life required an oxygen-free environment initially, then life produced oxygen in the atmosphere, and now animal life requires oxygen.

NOTES

1. Michael S. Turner, "Why Is the Temperature of the Universe 2.726 Kelvin?" *Science* 262 (5 November 1993): 861–66.

2. Barry Parker, *The Vindication of the Big Bang* (New York: Plenum Press, 1993), pp. 264–66.

3. Ed Tryon, "Is the Universe a Vacuum Fluctuation?" *Nature* 246 (14 December 1973): 396.

4. Thomas K. Gaisser, *Cosmic Rays and Particle Physics* (Cambridge: Cambridge University Press, 1990), p. 140.

5. P. A. Cox, *The Elements: Their Origin, Abundance, and Distribution* (Oxford: Oxford University Press, 1989), p. 179.

6. Stuart Ross Taylor, *Solar System Evolution: A New Perspective* (Cambridge: Cambridge University Press, 1992), p. 115.

7. Ibid., p. 23.

8. Ibid., p. 217.

9. Tjeerd H. van Andel, *New Views on an Old Planet: A History of Global Change*, 2d ed. (Cambridge: Cambridge University Press, 1994), p. 271.

10. Ibid., p. 281.

11. John J. W. Rogers, *A History of the Earth* (Cambridge: Cambridge University Press, 1993), p. 48.

12. Van Andel, *New Views on an Old Planet*, pp. 282, 305.

7

OUR COSMIC GENESIS II
The Origin and Advancement of Life

AT FIRST THOUGHT, LIVING ORGANISMS seem entirely different from nonliving matter. Animals can function as independent beings, making decisions and moving about under their own control. In spite of the apparent differences between animals and nonliving matter, the matter in organisms obeys the same laws that apply to nonliving matter. Thus, the physical laws and forces that were discussed in chapter 5 apply equally to living and to nonliving matter. At the level of atoms and molecules, matter within living bodies behaves the same way as matter within nonliving objects.

The obvious differences between living and nonliving matter are due to several causes. Chemically, living matter has high concentrations of water and carbon compounds. Carbon, because of the structure of its atoms, can produce an enormously diverse set of compounds with itself and with other chemical elements. This richness of carbon chemistry is rarely developed in nonliving matter, while life depends upon this abundance of complicated carbon compounds.

The main difference between living matter and nonliving matter, however, is in the elaborate *organization* of the material in the living organism. The wonderful organization of living matter is possible because of the great complexity of carbon chemistry. The organization has been developed by millions of years of experience and improvement, combined with an amazing method of passing this information from one generation to the next. Perhaps the most wonderful thing about life is that it is not the product of a purposeful design. Instead, it is the result of the normal operation of physical laws in a favorable environment. Let's begin our discussion of life by defining what we mean by living matter.

A DEFINITION OF "LIFE"

Life occurs in many forms, from microscopic bacteria to large plants and animals. Let's consider what all life has in common. This is easier than it might be because all life on earth appears to have a common ancestor and all living beings share many essential properties with this common ancestor. I will list some of these properties that are possessed by all organisms.

1. *All life can reproduce itself.* The reproduced organism carries information passed from its ancestors in the form of certain giant molecules. Copies are made of this information, and life may multiply if conditions are favorable. The copying of the genetic information and the formation of new organisms take place by ordinary chemical processes. The chemical procedures followed in the construction of a new organism are controlled by the genetic information transmitted from the parent or parents of the new organism.

2. *A flow of energy is required to carry out the chemical processes involved in life,* such as reproduction and the construction of cells. All forms of life have ways of obtaining energy. Plants obtain energy from sunlight and use this energy to build tissues, grow taller, and so on. Animals obtain energy by eating plants or other animals. More basically, animals obtain energy by destroying energy-containing chemicals built up by other organisms. (Not very noble, but it's a living.)

3. *The chemical processes of life are carried out in a protected environment.* (This may not have been a characteristic of the earliest life, but it is so prevalent now that it is included in this list.) These processes occur inside a small volume surrounded by a

membrane. The membrane and the material contained within it make up a *cell*. Cells vary greatly in size, but are usually quite small. Many organisms have only one cell, but some plants and animals contain extremely large numbers of cells. Cell membranes control the passage of materials into and out of the cells.

The three requirements listed above allow us to distinguish between living organisms and nonliving matter, at least for life as we know it on Earth. Of course, many organisms possess other remarkable properties. Life possesses numerous wonderful adaptations that allow it to exist in many different environments and with various specialized relationships to other life. This diversity is another basic property of life.

In the last 150 years, science has made spectacular progress in understanding life. The theory of biological evolution has enabled us to understand how life has obtained such remarkable adaptations and such spectacular diversity. Nature's "blueprints," which specify the properties of organisms, have been discovered. The processes by which these blueprints are used to construct and maintain organisms are being discovered. Many basic life processes are understood in detail. With so much progress, it is no longer sensible to think that life is beyond human understanding.

LIFE'S SELF-COPYING BLUEPRINTS

The intricate organization of the matter within organisms was formerly a mystery. Organisms seem to form themselves without the intelligent intervention that is required in our manufacturing processes. The offspring of most animals resemble their parents. An Austrian monk, Gregor Mendel (1822–1884), made progress in understanding such facts. In 1865 he found that the probability of inheriting certain features or *traits* followed certain laws, but that the particular traits passed on to an individual are a matter of chance. Mendel believed that traits are transmitted to a descendant by certain "discrete hereditary elements" (now called *genes*) contained within the reproductive cells of the parents.

In the 1880s it was learned that the genes are contained inside certain bodies within cells called *chromosomes*. In 1944 Oswald T. Avery and his associates found that the chromosomes are composed of giant molecules of deoxyribonucleic acid, more conveniently called *DNA*. DNA had a name for a long time before its detailed structure was known. The structure of DNA was found by the American biologist James D. Watson and the English physicist Francis H. C. Crick in 1953. Their discovery was based on their own model-building and other workers' chemical analyses and studies of DNA crystals using X-rays. The structure of DNA showed how it can serve as a blueprint for constructing new life.

The wonderful molecular structure of DNA is like a spiral staircase or twisted ladder. The ladder rotates one turn in about ten steps. The spiral shape of the ladder is the source of the DNA nickname, the double helix. Each side of the ladder is made of molecules of *ribose*, a kind of sugar, alternating with *phosphate* ions. (An ion is an atom or molecule with a net electrical charge. The phosphate ion consists of a phosphorus atom combined with four oxygen atoms, one of which has an extra electron that gives it a negative charge.)

The most interesting parts of the ladder are the "steps." Each step is made up of two pieces, called *bases*. Each base is bound to one side of the ladder and to the other base. There are four kinds of bases: *adenine*, *thymine*, *guanine*, and *cytosine*. They are conveniently represented by their initials: A, T, G, and C. If we know one base in a step, we can predict which base is its partner. The sizes and structures of the bases are such that A fits well with T and G fits well with C, and other combinations don't work. The AT and GC combinations have nearly the same length, so the separation between the ladder's sides is nearly the same for either combination.

Although the two bases in one step are always related to each other, the sequence of bases along one side of the ladder is unpredictable: there is no reliable pattern in the kinds of bases that follow each other along the ladder. Consequently, DNA is not one specific molecule. Instead, it is an enormous family of molecules that have various strings of bases. *The sequence of bases is meaningful, how-*

ever, and it is a coded message for how to construct an organism! DNA is a sort of tape that contains the blueprint for an organism.

Each cell contains DNA. In order to carry out the instructions encoded within DNA, a copy is made of the string of bases on one side of the DNA "ladder." This copy is called *ribonucleic acid*, or *RNA*. RNA is very similar to one side of the DNA ladder, except that where DNA contains thymine (T), RNA has a base called *uracil* (U). In complex cells such as in humans, the RNA moves from where it is formed at the DNA in the cell's *nucleus* to another part of the cell where there are *ribosomes*.

A ribosome has been described as a "catalytic assembly bench" for *proteins*.[1] It is a molecule that uses the encoded information that has been transcribed onto the RNA "tape" to bring together *amino acids* to form proteins. (A protein is a molecule made up of a string of subunits called amino acids.) Living organisms "make do" with twenty amino acids, a small fraction of all known amino acids. Particular sequences of three bases in the RNA correspond to specific amino acids. For example, CUU signals that the amino acid leucine should be added to the protein. Another three-base sequence signals that the construction of a protein is finished.

You may ask why assembling proteins is so important. Proteins carry out most of life's processes. They are used to construct living tissue. Muscles operate by the contraction of protein fibers. Hemoglobin is a protein that carries oxygen in the red blood cells. An important role for proteins is as *enzymes*. Enzymes enable life's chemical processes to occur. The enzymes accomplish their work in promoting chemical reactions by bringing other chemicals together, but the enzymes themselves are not used up in these processes. Enzymes can assemble specific useful molecules, either proteins or nonproteins, and they can do this rapidly and reliably.

In the previous paragraphs I have traced the "chain of command" by which DNA controls the chemistry of life. This shows how DNA is a blueprint for life. The chain of command can be represented by DNA → RNA → protein. Sometimes the protein is not the final product, but is an enzyme that makes other chemical reactions happen.

The DNA in a cell comes from a parent cell. Cells reproduce by dividing into two cells. When this happens the DNA needs to be copied, so there is a full set of DNA for each cell. This is done with the help of enzymes previously produced in the parent cell. The enzymes assist in separating the two sides of the DNA molecule, like opening a zipper. Then other enzymes connect nucleotides to the two pieces, accurately matching appropriate base pairs, such as A with T and G with C. In this process, each half of the original DNA serves as a template for the construction of a DNA molecule. In the end there are two sets of DNA, one for each cell. *Thus, DNA organizes all the matter in a cell and also arranges for copying itself in order to make new cells.*

In a later section, it will be argued that all life has a common ancestor. If the process of making perfect copies of cells that are identical to the parent cells were all that happens in nature, then only one kind of life would exist on Earth. There would only be perfect copies of a very primitive ancestor. But the "perfect copy picture" is too simple. Some cells in higher animals have a built-in sensitivity to their environment, so their development depends on conditions. This allows cells found in certain organs of the body to become very different from cells in other parts of the body. DNA can also change, due to imperfect copying or due to effects of certain chemicals and radiation. These DNA changes, called *mutations*, have generated the great diversity and complexity we see in the life around us. This happened by a long and painful process in which most mutations were harmful, but a small fraction of the changes were beneficial.

CHARLES DARWIN AND NATURAL SELECTION

From 1831 until 1836, Charles Darwin (1809–1892) served as an unpaid naturalist aboard a British survey ship, the *Beagle*, on a round-the-world voyage. Before he was offered a position on this expedition, the young Englishman had planned to become a clergyman. Darwin's experiences on the *Beagle* changed the course of his life. Darwin took notes and collected specimens and fossils along the coasts of South America and in the Galapagos Islands west of South America, among other places. He attempted to understand the geology and natural history of the areas he visited.

With his religious background, Darwin expected that his observations would support creation as depicted in Genesis in the Old Testament.[2] In this picture, each kind of animal had been created directly by God. It was thought that the marvelous abilities of animals to survive were due to God's foresight. *The species of animals were thought to be unchanging and ideally designed for their environments.*

When Darwin visited the Galapagos Islands, he found that they resembled the Cape de Verde Islands near Africa in size and height, the volcanic nature of the soil, and the climate. The situations suggested that the two sets of islands should have similar animals, if the animals had been created to fit into their very similar environments. Darwin observed, however, that the animals and plants of the Galapagos were so similar to those of South America that every one "bears the unmistakable stamp of the American continent."[3] Darwin also noted that the animals and plants of the Cape de Verde Islands are closely related to those of Africa.

The Galapagos species were *not the same* as those in South America. Of twenty-six land birds in the Galapagos, twenty-five were distinct species from those found elsewhere. "Yet the close affinity of most of these birds to American species in every character, in their habits, gestures, and tones of voice, was manifest," according to Darwin.[4] The situation was similar for other animals and plants. To Darwin, if not for the modern reader, it was puzzling that the Galapagos species were so similar to species from the nearby continent. He reasoned that the islands did not resemble South America in geology, climate, or other conditions, so the animals created to live on the islands should not have resembled those that had been created to live on the continent.

Darwin noticed that species of animals living on the different Galapagos Islands were similar but somewhat different. He had also noticed a succession of closely related species as he traveled south along the eastern coast of South America. There were also great similarities between fossils of extinct animals in South America and the living animals. All of these facts pointed toward a sort of continuous variation of some animals that were near each other in space and in time. He came to suspect that animal and plant species were slowly changing over large times and across large distances or when separated by physical barriers, such as mountain ranges or bodies of water. He knew that individuals of a species differed somewhat from each other and that these differences could be inherited. This could allow species to change, but at first he could not understand how species could change but still remain so well adapted to their environments.

In 1837, after completing his journey, Darwin started keeping notebooks to collect evidence to support and improve his ideas. In 1838 he read the gloomy *Essay on the Principles of Population* by Thomas Robert Malthus (1766–1834). Malthus argued that human populations tend to increase by a *large fraction* in each generation. If left unchecked, this would result in an enormous growth of the population while the production of crops increases more slowly. The resulting food shortages are only checked by population losses from disasters such as war, famine, and plague. To Malthus, human life is a constant struggle for survival and only the fittest survive.

Darwin knew that animal and plant populations are capable of enormous growth. He knew that in most species only a small fraction of offspring live long enough to reproduce. It struck him that the differences between individuals give some animals (and plants) an advantage in this struggle. More individuals with useful qualities or favorable variations survive and reproduce, passing the favorable traits to their offspring. By this process, favorable traits become more common and unfavorable variations become rare within the population. This process explains how life remains well adapted in a changing environment. Darwin called this process *natural selection*.

By natural selection, a species could adapt to a new environment. A species that reached an isolated habitat (such as the Galapagos Islands) would gradually and automatically undergo changes to adapt to the new conditions. *This could produce a new species.*

The different environments on nearby islands could even produce a number of species, as had been observed by Darwin.

Darwin was very slow to publish his ideas. In 1858 he was surprised by a letter from Alfred Russel Wallace (1823–1913), another English naturalist. Darwin learned that *Wallace had also studied species on islands and had also hit upon the process of natural selection after reading Malthus's book*. Both men agreed to publish short papers on their theories later that year. In the next year, 1859, Darwin published his great work *On the Origin of the Species by Means of Natural Selection, or the Preservation of Favoured Races in the Struggle for Life*. This book, supporting evolution, was revolutionary in its effect on our understanding of life and humanity.

EVOLUTION AND LIFE

The implications of Darwin's ideas went far beyond the issues which Darwin discussed in his book on the origin of species. Perhaps Darwin realized this but did not want to face the wrath of the great majority of Christians who, at that time, took the book of Genesis literally. Natural selection leads to the scientific theory of *biological evolution*. By applying the idea of natural selection, it is easy to understand how new species could form. Fossils show how species actually developed from their ancestors. Related species share common ancestors.

In the next section it will be argued that all life-forms (plants, animals, fungi, and one-celled organisms) share a common ancestor. This implies that family-tree diagrams for all living organisms could, in principle, be drawn all the way back to this common ancestor. The process of natural selection helps to explain why life is so diversified, with each species so well adapted to its environment. New species are sometimes more complex than their ancestors. The origin of species by natural selection allows for the great complexity present in some organisms.

Fossil collections are much more complete now than in Darwin's time. If you have seen the sequences of fossils showing the development of the horse, you probably have a feeling for the reality of evolution. The fossils are convincing evidence that evolution has occurred through millions of years. No large modern animals have fossils among the oldest layers of rocks. Such animals have developed in the last 600 million years or so, relatively recently in the earth's history.

During Darwin's time little was known about genetics, the science dealing with inherited traits. (Mendel's work was largely ignored before 1900.) Today, it's known that *mutations* are changes within DNA that result in new characteristics in an organism's descendants. The variations between members of a species that are assumed in Darwin's theory are produced by mutations. Mutations are often caused by replacement of a single nucleotide by another nucleotide in the DNA of an organism's reproductive cell. This tiny error can produce major changes in the descendants. Usually these changes are bad, and the descendants reproduce less frequently than normal for the species. The bad segment of DNA tends to become rare or extinct within the population. Occasionally, a mutation is favorable and leads to an improved ability of the organism to survive and reproduce. The good segment of DNA tends to become common within the species.

Darwin's theory of evolution applies to the history of humanity, of course, just as it applies to all organisms. Humans and chimpanzees have a common ancestor. To human eyes this ancestor would look apelike. To a chimpanzee, this ancestor would probably look "humanlike." If you go back far enough, we have ancestors that were *much* more primitive and less intelligent than apes. So why complain about apes in the family tree?

Since I am discussing evolution, I should mention the creationist view. To deny evolution and support creationism is not science. It is an act of religious belief. Evolution is a theory that essentially all biologists and geologists support. Since the evidence supports it so strongly, denying evolution comes close to demonstrating ignorance of the subject, while asserting an opinion. This stance is granted no respect by those who have serious scientific training or interests. As a theological position, the denial of evolution seems to amount to declaring the "Creator" guilty of colossal deception, since the evidence supporting evolution has been set into billions of tons of stone.

ONE ANCESTOR OF ALL LIFE

Is there a common ancestor of beets, bats, and bacteria? The answer seems to be yes. All life uses the DNA→RNA→protein mechanism. The DNA and RNA of all life have the same general structure and use the same bases, with a few exceptions. The same twenty amino acids are used in constructing the proteins of all life, while over one hundred amino acids occur in certain types of meteorites.[5] Thus, all proteins in living organisms are constructed of twenty "standard parts" out of a much larger number of possible parts.

The genetic code described earlier, in which a sequence of three bases specifies the use of a particular amino acid in constructing a protein, is nearly identical in all life found on Earth. The identification of the amino acids in all forms of life by the same set of *twenty arbitrary three-letter codes* (with choices of A, T, G, and C) is *extremely* unlikely to be an accidental coincidence.

Another similarity of all life is that the processes that require energy use the same energy source, a chemical called *adenosine triphosphate* (ATP). When plants acquire energy from sunlight or when animals obtain energy from food molecules, the energy is used to construct ATP. The ATP is then used as the energy source in most cellular processes. ATP is to life as money is to commerce. It is the "common currency" for paying energy "debts." The fact that all life uses ATP for this purpose is additional evidence that all life has a common origin.

But what was the common ancestor of all life? No one knows, but it was probably a form of bacteria. These are the simplest forms of life and are present in the earliest fossils. The ancestral organisms were probably similar in some ways to certain modern bacteria that live in the absence of oxygen. They may have lived in hot water, since some primitive bacteria still live in hot water and the early earth is thought to have been quite hot. Whatever the ancestral organism was like, *the fact that there was one* implies that each person's family tree starts with this organism. Each person, and every other living being, has *an unbroken chain of ancestors going back billions of years* to a primitive, one-celled organism.

THE ANTIQUITY OF LIFE

Evidence exists that life formed quite early in the earth's history. Remember that the earth is 4.6 billion years old. Until 4.1 billion years ago the surface of the earth was bombarded so heavily by planetesimals that life was probably impossible.[6] The earliest suggestion of life comes from the oldest known sedimentary rocks. These rocks are 3.8 billion years old and occur in southwestern Greenland. Carbon within these rocks contains unusually high amounts of the isotope carbon–12 compared to carbon–13. Organisms growing by absorption of sunlight can produce this effect, which is hard to explain otherwise. This suggests, but does not prove, that life already existed 3.8 billion years ago.

More certain identification of early life comes from somewhat "younger" rocks. Certain large-scale colonies of bacteria produce finely layered rocky bodies called *stromatolites*. Stromatolite fossils are present in 3.5-billion-year-old rock from the "North Pole" area of Western Australia and from the Barberton Mountains in Swaziland in southern Africa.[7] Living stromatolites can still be found in very salty waters such as in the Hamelin Pool in Shark Bay on the west coast of Australia and at Abu Dhabi on the Arabian Peninsula.[8]

Besides indirect evidence from stromatolites, more direct evidence of early life comes from fossil imprints of bacteria found in 3.5-billion-year-old rocks. Some fossils of bacteria are found in stromatolites,[9] strengthening the interpretation of the stromatolites as fossil bacteria colonies. From this evidence, the origin of life on Earth probably occurred between 3.5 and 4.1 billion years ago, relatively early in the time interval in which environmental conditions allowed it to form.

THE ORIGIN OF LIFE

We can learn how life's processes work by studying living or pre-served organisms. Scientists can learn how life diversified by "reading" the fossil record of life. It is relatively difficult, however, to learn *how the first life on Earth began* because the first life is not available at present and it probably did not leave fossils. Consequently, the subject of the origin of life is much more speculative than most other areas of biology. A number of different and even contradictory ideas exist about how life started. Because the evidence is so scanty, none of these ideas can be ruled out. Only some basic facts and a few ideas concerning the origin of life will be discussed here.

Earth's atmosphere did not contain molecular oxygen (as opposed to oxygen bound in carbon dioxide molecules) at the time of the origin of life. *This lack of oxygen was probably necessary for the formation of life*, since an abundance of free oxygen would have prevented the synthesis of many of the compounds necessary for life's origin.[10] By definition, the *first life* developed in the absence of other life. Thus, there were no predators or competitors for food. Since there were no living organisms to "spoil" compounds useful to life, the compounds built up in the oceans in a way that is now impossible. The oceans became a sort of weak soup that did not need to be "canned" for preservation because of the absence of life or free oxygen. The first life may have been *some trivial little soup-spoiling organism*.

Experiments have been done which show that many substances needed for the origin of life would have been present early in the earth's history. A classic study, the Urey-Miller experiment,[11] used artificial lightning (electrical discharges) in a flask containing a mixture of substances. This setup produced some amino acids and other organic compounds necessary for life. The experiment was probably not realistic since the gas mixture contained much more hydrogen than is now thought to have been present in the early earth's atmosphere. However, later experiments with more realistic conditions have also yielded the basic chemicals for the origin of life.

The observation of amino acids and other organic compounds in meteorites has intrigued some scientists dealing with the origin of life. The early earth may have been enriched in organic chemicals by meteorites and comets that reached the earth's surface. Still, it is a long way from the basic chemicals of life to a living system. DNA probably did not just "fall together" out of a "primeval soup" of organic chemicals. There may have been a long history of evolution of nonliving chemical systems before systems developed that fit our definition of life. Such systems may have increased in complexity by many small steps, without taking any very difficult steps.

At its origin, life was simpler than it is now. One way it may have been simpler was in the DNA→RNA→protein sequence that was discussed earlier. The DNA part may have been left out. RNA can carry the information needed to produce proteins. Even now, some viruses reproduce by copying RNA, without using DNA at all. The idea that life originated with RNA, but without DNA, is called the "RNA world" hypothesis.[12] It was probably simpler for the first life to form in an RNA world rather than by the more complicated process that uses both DNA and RNA.

Even with the presence of the basic chemicals, the origin of life still required an energy source and certain catalysts or enzymes. A source of energy was needed to allow complex chemical reactions to occur which built up large biological molecules. These processes required many steps, and some steps required energy. An energy source made it possible for the entire process to go forward.

There are many conceivable energy sources for the first life. One possibility is heat from hot springs on the sea floor, called *hydrothermal vents*, that occur where the ocean's floor is spreading. Hot springs on land, volcanoes, and lightning are other possible energy sources. During life's beginning, the lack of an atmospheric ozone layer allowed intense ultraviolet radiation from the Sun to reach Earth's surface. Iron ions in seawater could have absorbed energy from this radiation. These ions could have supplied energy for developing life, according to one idea.[13]

A number of natural catalysts have been identified as potential aids to the formation of the first life. These include clays, metals, and iron pyrite (fool's gold). One kind of clay, called *montmorillonite*, can form small RNA-like molecular structures from the basic nucleotides. Clays can also aid the formation of amino acids at temperatures well above the boiling temperature of water. The early chemistry of life may also have been aided by organic catalysts or *enzymes*. Enzymes speed up certain chemical processes connected with life, although the enzymes themselves are not used up in the processes. The sterile, oxygen-free conditions may have allowed significant concentrations of enzymes to build up before life existed.

Life, as defined in this chapter, requires cell membranes or cell walls. Perhaps the earliest replicating organic systems did not have membranes, but eventually organisms obtained them. Cell membranes are composed of proteins and fatty materials called *lipids*. How these cell boundaries originated is a basic, difficult question concerning the origin of life.

A wide variety of organic chemicals have been detected in clouds of dust and gas between the stars. Meteorites and comets also contain many compounds useful to life. It has been proposed that life began in space rather than on Earth. This is a minority view, but no overwhelming evidence or arguments have eliminated this possibility.

SOLAR-POWERED LIFE

Many of the most important life processes arose in the very long period between the origin of life and the much more recent development of multicellular plants and animals, which started "only" about 600 million years ago. Let's discuss an extremely important process: obtaining energy. Life uses three methods to obtain energy: *fermentation, photosynthesis,* and *respiration*. All three processes were probably developed by bacteria before more complicated life existed.

The earliest life may have been able to live by fermenting energy-rich organic chemicals produced by solar energy or other sources. In *fermentation*, chemical compounds are converted by organisms into less energy-rich compounds, giving energy to the organism. This energy is used to construct ATP, the chemical that is the main energy carrier within cells. Many bacteria and yeasts use fermentation to obtain energy. Fermentation can occur without oxygen and is used by plants and animals when oxygen is not available. Fermentation is involved in the production of wine, vinegar, beer, yogurt, and cheese, and in the rising of bread dough.

As an energy source for life, fermentation has some disadvantages. A species that depends on fermentation often requires fairly specific organic chemicals as food. Limited amounts of these chemicals may be available, thus limiting the population of the dependent species. Fermentation is not efficient in extracting all of the food's chemical energy. In fact, the waste products of one fermenting organism can often be used as food by another organism in *its* fermentation process.

Photosynthesis, in which sunlight is the energy source, is a more dependable way of obtaining energy than fermentation. Photosynthesis is thought to have been employed by bacteria that lived in the 3.5-billion-year-old stromatolites that form the earliest clear evidence of life on earth.[14] With the development of photosynthesis, life tapped a steady source of energy from the Sun. *Life used this energy to produce its own food chemicals*, such as *glucose*, a form of sugar.

Carbon dioxide served as an abundant source of carbon and oxygen in the production of glucose by photosynthesis. The source of the *hydrogen* in glucose was more of a problem. Some bacteria can obtain hydrogen from the compound hydrogen sulfide. Others can obtain it from organic molecules such as ethanol, the active ingredient in alcoholic drinks. These chemicals are not always abundant. Water is a much more abundant and reliable hydrogen source. Bacteria evolved processes to use water as a source of hydrogen, producing *oxygen* as a waste product. But this produced a global crisis for life, since *oxygen was toxic to the earliest life*.

To us, as modern organisms, it seems ironic that oxygen was originally a toxic chemical waste product that nearly destroyed all life by polluting the atmosphere and the oceans! Oxygen is a very active chemical element, and it can alter many of life's essential chemicals. Modern life-forms have adapted to the presence of oxygen by producing various enzymes that block oxygen's harmful effects, converting dangerous chemicals produced by oxygen into harmless compounds.

The very early forms of life, however, had evolved in the absence of free oxygen. Most of them did not possess protective enzymes, and could be killed by oxygen. The bacteria that produced the oxygen must have had protection against oxygen since they had developed processes in which they were constantly exposed to it. So they could happily proceed with their new chemical processes while polluting the world and thereby killing their competitors! Other forms of bacteria could perish, adapt to oxygen, or live in oxygen-free environments, as some still do.

The bacteria that used photosynthesis and produced oxygen became enormously successful, since sunlight and water, their most basic needs, are very abundant. At first the oxygen did not remain in the atmosphere, but combined with iron, sulfur, or other materials. Eventually oxygen replaced most of the previously abundant carbon dioxide in the atmosphere. The earth's iron deposits give evidence of a change from low oxygen to more abundant oxygen about two billion years ago. *Except for the first life, the humble photosynthesizing bacteria, often called* blue-green algae, *changed the earth more profoundly than any other organism.*

With abundant oxygen, *respiration* became possible and was taken up by many organisms. In this process, all the carbon and hydrogen in such food molecules as glucose are combined with oxygen, producing water and carbon dioxide. This is very efficient in extracting all the chemical energy from the food. For example, from one molecule of sugar, *fermentation* by bacteria yields two ATP molecules, while *respiration* yields up to thirty-six ATP molecules.[15] Respiration can be used by photosynthesizing organisms to obtain energy from stored glucose when it is dark. Organisms that do not produce their own food can also use the respiration process to "burn" their food efficiently. Fermentation, photosynthesis, and respiration are still vital to modern plant and animal life, billions of years after the development of these processes by bacteria.

BEYOND BACTERIA

For about the first half of the time that life has lived on Earth, bacteria and some other rather exotic one-celled creatures called *archaea* were the only forms of life.[16] In bacteria, the cell structure is relatively simple, with a circular strand of DNA spread around within the cell, where it is free-floating. Organisms with this kind of cell are called *prokaryotes*. They are all very small one-celled organisms that reproduce by the relatively simple processes of cell division.

Except for bacteria and archaea, all forms of life are *eukaryotes*. They have cells that are much more organized and are usually larger than the cells of the prokaryotes. Eukaryote cells all have a nucleus. The nucleus is a region inside the cell where the cell's DNA is enclosed in a membrane, just as the entire cell is enclosed within another membrane or cell wall. Most eukaryotes also have other "parts" in their cells, called *organelles*, that are enclosed by membranes. Examples of organelles are *mitochondria* and *chloroplasts*. *Mitochondria* are organelles where energy is extracted from food molecules and ATP is produced for use in other parts of the cell. *Chloroplasts* are organelles in which photosynthesizing organisms use chlorophyll to absorb energy from sunlight.

Eukaryotes make up the following major divisions, or *kingdoms* of life: *protista*, fungi, animals, and plants. The protista contain the simplest eukaryotes, such as the amoeba, a tiny, one-celled organism that can move about in a simple bloblike fashion. According to fossil evidence and the detection of chemicals called *steranes* in ancient rock, some eukaryotes were present by about 1.7 billion years ago and may have originated earlier. However, many other groups of eukaryotes arose in a relatively short period of time about 1.1 billion years ago. Eukaryotes were the first organisms capable of feeding on other organisms by enclosing them. They also have a much more complicated process of handling their genetic material when cell division occurs. This may have been necessary because *eukaryotes carry much more DNA than do prokaryotes*.

Prokaryotes and eukaryotes seem to represent different

approaches to life. Comparing them is like comparing a hiker with a backpack to a traveler with an automobile and a mobile home. The prokaryotes are enormously successful using a simple approach to life. Eukaryotes have gone toward great complexity and are also very successful. For our purposes, eukaryotes are more interesting, since free will and the soul do not seem to be relevant issues with the prokaryotes.

Unlike the prokaryotes, many eukaryotes reproduce sexually. This means that each individual has two complete sets of DNA: one set from each parent. When cell division occurs, the DNA and various proteins form separate rodlike pieces called *chromosomes*. Each human cell, except for the reproductive cells, contains DNA that is divided into forty-six chromosomes, that is, *twenty-three pairs of corresponding chromosomes*. The mother's egg cell and the father's sperm cell each contained twenty-three chromosomes, giving the resulting individual a total of forty-six chromosomes.

Each chromosome pair consists of one chromosome from the mother and one from the father. The choice of *which chromosome is inherited from the pair carried by the parent* is made randomly. For example, your mother might have given you *either of her two copies* of chromosome six: the one from her mother or the one from her father. Which copy you receive is a matter of chance. With forty-six such random choices, it is not surprising that the set of genes of each individual is unique, except for the special case of identical twins.

Actually, sexual reproduction is not quite as simple as described above, since *crossing over* can occur. Consider the example of receiving chromosome six from your mother. By the process of crossing over you could receive, say, 40 percent of chromosome six from your maternal grandmother and the remaining 60 percent of chromosome six from your maternal grandfather.

Sexual reproduction is so widespread that it must have important advantages over a simpler system of passing on a single "best" set of genes found by evolution. For complicated organisms that may live in various environments, nature seems to be unable to find a single best set of genes. By sexual reproduction there is a

great variety of combinations of genes present in the various individuals of the total population of a species. There are probably
some individuals among a population who will manage to survive
a flood, famine, fever, or frost, although no individual may be able
to survive all of these. Genetic diversity probably gives a species an
improved ability to adapt to changing environments and to differing conditions in various parts of its range.

The first eukaryotes and the first sexually reproducing organisms were probably one-celled organisms. Multicellular animals, or
metazoa, made relatively late appearances in the history of life. The
oldest abundant fossils come from the *Cambrian* period of geological history, which lasted from about 545 to 510 million years ago.[17]
At that time animals developed with hard parts such as shells and
spines. Unlike the softer parts, these structures form fossils frequently. Most of the basic body-plans for animals appeared within
the Cambrian period. This rapid increase in the diversity of animals
has been called the "Cambrian Explosion."

An obvious question is whether many soft-bodied metazoans
lived *before* the Cambrian, but did not produce fossils. Actually
there are various fossils of soft-bodied animals, called the *Edi-
acaran fauna*, which lived between about 580 million years ago
and the beginning of the Cambrian. They were first found in the
Ediacara Hills in the interior of Australia, and later were found on
all continents except South America and Antarctica. Some of these
fossils were quite large. Tracks of burrowing and crawling organisms are also found from this time period.

Convincing examples of metazoan fossils or tracks of animals
have not been found from times earlier than 580 million years ago.
The fossil record suggests that the metazoans originated about six
or seven hundred million years ago. It is difficult, however, to rule
out earlier dates since the very early metazoans were probably
small and soft-bodied, and may not have left many fossils.

The timing of the origin of life and of the metazoans is interesting. *The origin of life occurred relatively quickly once conditions
on Earth allowed it. But the metazoans did not develop until the last
20 percent or so of the time since the origin of life.* Metazoans are

much more complex than the prokaryotes and require large amounts of DNA to control their development. Because of this, their origin probably had to wait until eukaryotes developed, at about 50 percent of the time since the origin of life. However, why didn't metazoans develop shortly after the eukaryotes formed?

The metazoans require higher levels of atmospheric oxygen than one-celled organisms. Low oxygen levels may have prevented multicellular life from developing until about seven hundred million years ago. *It is also possible that the development of metazoans was more difficult for nature than the origin of life.* Advanced animals are more than colonies of cells, since the cells are organized into many different kinds of tissue. Metazoans require DNA that carries instructions for the production of the entire organism, not just the development of a single cell. This is a tricky process, requiring the orchestration of an embryo's development. *Perhaps it was even more difficult for evolution to find ways to do this than it was for the first life to develop from lifeless chemicals.*

MUSCLES: ORGANS SPECIALIZING IN MOVEMENTS

One of the most mystifying properties of animals is their ability to move themselves. Even in modern times this seems mysterious, because the structures which animals use to move are so different from the structures of our mechanical inventions that move themselves. Our vehicles move by ways that are well understood while animals, including ourselves, move in ways that are hard to explain in a fundamental way. We say that animals move by the contraction and relaxation of muscles, using chemical energy obtained from food. But how do the muscles contract? Can this be explained in terms of atoms and molecules?

In order to get a basic understanding of how animals move, we must consider the molecular structure of muscle tissue. As with many other problems, life uses *proteins* to solve the problem of self-propulsion at the molecular level. In spite of the difficulty in learning about such extremely small structures, much has been learned about how muscles work.[18] The structure of a human *skeletal muscle* is shown in the figure, and its functioning is described below.

The muscle drawn in the figure is made of cylindrical fibers. Within these fibers are smaller cylindrical structures called *myofibrils*. The myofibrils, shown about five thousand times actual size, show a pattern of dark regions (*A bands*) and light regions (*I bands*). The bands are related to the structure of the myofibril, blown up below the center of the figure. The dark bands are due to relatively thick, knobby fibers called *myosin filaments*. These filaments are seen (lower right) to be bundles of sprout-shaped molecules of the protein *myosin*, shown about two hundred thousand times actual size. Although the myosin molecule is tiny, it still contains tens of thousands of atoms. The "knobs" on the myosin filaments are the pairs of "heads" at the end of myosin molecules.

Parallel with the myosin filaments are thin *actin filaments*, partly made up of a pair of twisted strings of globular *actin molecules*. The actin filaments are interleaved with the myosin filaments. The I bands are regions where there are only actin filaments. The muscle contracts by increasing the overlap of the two kinds of filaments. When the muscle relaxes, there is less overlap of the filaments.

Now let's see what actually *moves* the muscle. It's not clear from the figure, but the myosin molecule heads (the knobs on the myosin filaments) can reach out and contact the actin filament. Such a knob then attaches itself to a globule in the actin filament. Next an ATP molecule also attaches to the knob, giving the myosin molecule energy to take the following steps: (1) the knob releases the actin globule, (2) the myosin molecule changes its shape so that the knob shifts forward to the next actin globule and attaches to it, and (3) the myosin molecule snaps back to its original shape, which pulls the actin filament, producing greater overlap between the actin and myosin filaments. After step 3, the knob is ready to go back to step 1 and repeat the cycle.

The knobs all along the overlapping filaments participate in the

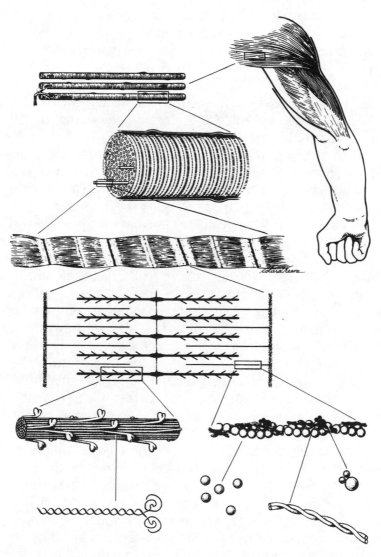

A series of magnified images shows the structure of muscles from the human scale down to the molecular scale. Muscle movements result from interactions between the sprout-shaped myosin molecules and the actin filaments, part of which resemble a twisted pair of strings of beads. The process is described in the text. From *Concise Histology* by Fawcett and Jensh, 1997, Chapman and Hall, Fig. 9–9. Reprinted by permission.

steps described above. *This activity contracts the muscle.* For each cycle of the steps, a knob advances one actin globule and an ATP molecule is converted into an *ADP molecule* that has less available energy. The knob shifts only about 5×10^{-9} meters per cycle, so about five million cycles are required to move an inch. Muscles have very large numbers of these units connected end-to-end, and all the units can be shortening at the same time, so that the muscle can contract rapidly.

Muscles are often relaxed, and this implies that the processes described above can be prevented from occurring. Two kinds of proteins not shown in the figure are involved in controlling the muscle contraction.[19] They are attached to the actin and are in contact with each other. When calcium ions are not present within a muscle, one of the proteins, called *troponin*, has a shape that allows a second kind of protein, called *tropomysin*, to position itself so that it prevents the knobs from attaching to the actin globules. As a result, the muscle does not contract.

Motor nerves from the central nervous system control when the muscles contract. A signal to contract is carried to a number of muscle fibers by multiple branches near the end of each nerve cell. Thus, a single nerve cell (also called a *neuron*) controls the contraction of many fibers and many *thousands* of myofibrils. The neuron's signal causes calcium ions to be released within the muscle. The calcium ions combine with the troponin and change the troponin's shape. This shape change of the troponin moves the tropomysin so that the knobs are no longer blocked from attaching to the actin globules. This allows muscle contraction to occur.

When a detailed discussion of muscle contraction is given, it all sounds very mechanical. It is. The more interesting parts of the actions of a person, for example, are the operations of the brain and other parts of the nervous system. Nerve signals generated by the nervous system control muscular contractions, producing a person's actions. The following chapters will deal more with these seemingly less mechanical aspects of our existence.

NOTES

1. Christian de Duve, *Vital Dust* (New York: BasicBooks, 1995), p. 66.

2. Alan Moorehead, *Darwin and the Beagle* (Harmondsworth, England: Penguin Books, 1971), p. 37.

3. Charles Darwin, *The Origin of Species*, ed. J. W. Burrow (Harmondsworth, England: Penguin Books, 1968), p. 385.

4. Ibid., p. 385.

5. NASA, *Search for the Universal Ancestors*, ed. H. Hartman, J. G. Lawless, and P. Morrison (Washington, D.C.: NASA Scientific and Technical Information Branch, 1985), p. 78.

6. Simon Conway Morris, *Understanding the Earth*, ed. Geoff Brown, Chris Hawkesworth, and Chris Wilson (Cambridge: Cambridge University Press, 1992), p. 438.

7. Ibid., p. 443.

8. Stjepko Golubic, *Environmental Evolution*, ed. Lynn Margulis and Lorraine Olendzenski (Cambridge: MIT Press, 1992), p. 133.

9. de Duve, *Vital Dust*, p. 5.

10. Tjeerd H. Van Andel, *New Views on an Old Planet* (Cambridge: Cambridge University Press, 1994), p. 282.

11. The experiment was performed in 1953 by Stanley L. Miller, a student of Harold C. Urey at the University of Chicago. The experiment is discussed by L. E. Orgel in *The Origins of Life: Molecules and Natural Selection* (New York: John Wiley & Sons, 1973), pp. 124–30.

12. Leslie E. Orgel, "The Origin of Life on the Earth," *Scientific American* 271 (October 1994): 78.

13. de Duve, *Vital Dust*, p. 43.

14. Lynn Margulis, *Early Life* (Boston: Jones and Bartlett, 1984), p. 66.

15. Ibid., p. 67.

16. Virginia Morell, "Life's Last Domain," *Science* 273 (23 August 1996): 1043–45.

17. S. A. Bowring et al., "Calibrating Rates of Early Cambrian Evolution," *Science* 261 (3 September 1993): 1293–98.

18. Ivan Rayment et al., "Structure of the Actin-Myosin Complex and Its Implications for Muscle Contraction," *Science* 261 (2 July 1993): 58–65.

19. R. D. Keynes and D. J. Aidley, *Nerve and Muscle* (Cambridge: Cambridge University Press, 1991), p. 159.

8

ADAM'S TOOL KIT
The Evolution of Human Abilities

THERE IS A TENDENCY TO view human talents as not only above the level of other animals found in nature, but even *beyond the reach of nature*. This leads to a belief that we are partly supernatural beings. That belief will be discussed in the following chapters. The present chapter shows how our abilities, including some of our uniquely human talents, arose by entirely natural processes.

Although early humans evolved from animal ancestors, we have become unquestionably the most dominant large organisms on Earth. In this chapter I will describe how humanity has acquired such supremacy. Our special talents have evolved from abilities that are also present to some extent in other animals. The capabilities of some other animals are also quite impressive. The whole picture does not suggest that humans are essentially different kinds of beings than other animals. Instead, humans are the result of a very successful and very extensive improvement of abilities that are also present in some other animals.

In many ways we are *not* outstanding among animals. Many animals surpass us in size, strength, length of life, fast running, keen vision, smell, or hearing. We excel, however, in making and using tools, control of our surroundings, and planning and communicating. Our exceptional talents are mostly related to our brains. Our brains are products of a particular evolutionary path that has resulted, after millions of years of progress, in our outstanding abilities.

CONTROL OF MOVEMENTS BY NERVOUS SYSTEMS

Multicellular animals obtain energy-rich chemicals (food) by eating other organisms. Food is obtained by movements to capture and swallow the food organisms. These movements occur by changes in the lengths of muscle fibers. However, the movements need to be *controlled* in order to be effective. A predatory animal needs to approach its prey and bite at the right time. Even a jellyfish needs to coordinate its contractions to move through the water.

Animal movements are controlled by the combined activities of *sense organs* and *nervous systems*. Sense organs evolved to respond to specific facts about the animal's environment. A sense organ's sensitive cells, called *receptors*, interact with matter or energy in the animal's surroundings or body and produce electrical or chemical changes. These changes signal the presence of the matter or energy being detected. Some receptors respond to the organism's surroundings, such as sound receptors. Other receptors detect the internal status of the animal, such as the positions of the animal's joints.

Sensory neurons convey information about the animal's environment to other neurons that may "decide" to respond by "taking action." The words "decide" and "taking action" describe electrical and chemical processes in which strong sensory signals cause other neurons to produce certain signals. There may be a decision to contract certain muscles. In that case signals are sent to the appropriate muscles by motor neurons, producing muscular contractions.

The decision to take action may be made in an extremely complicated network of neurons, such as a person's brain. In some cases the decision is trivial. For example, the knee-jerk reflex is normally automatic. When certain tendons near the knee are tapped, receptors detect this and send a signal directly to motor nerves that cause muscles to contract, producing a small movement of the leg. This response is not affected by what we hear or see, or what we want to do, because it does not involve neurons that carry any information from the ears, eyes, or brain.

NEURONS, MEANING, PATTERNS, AND CATEGORIES

As implied above, a *neuron* is a kind of cell that specializes in communicating information and making decisions. There are many different types of neurons, but they have some common features. Many neurons have a shape resembling a microscopic uprooted tree. There is a branching system of thousands of roots called *dendrites*. They carry information *to the neuron* from contacts with other neurons. Dendrites lead into the cell body, analogous to the base of the tree's trunk. The cell body carries out many vital processes of the cell and it contains the cell's nucleus.

Coming out of the cell body there is often a long, very thin trunk, which is called an *axon*. The axon usually leads to a complicated system of branches. At the ends of each branch there is a small bulb-shaped enlargement called the *axon terminal*. This connects (*across a gap*) to another neuron or to a muscle fiber. The junction of the axon terminal and the other cell is called a *synapse*. The axon and its terminals are the output parts of the neuron. They carry signals away from the cell body to the synapses. At a synapse, the axon terminal is often connected to the dendrite of another neuron, which continues to process the nerve signals.

Let's take a brief look at how a neuron functions. Neurons often receive input signals at a synapse on a dendrite. The signal is a change of voltage of about one-tenth of a volt that lasts for about one-thousandth of a second. Such a temporary voltage change, which a physical scientist might call a *pulse*, is commonly called an *impulse* by physiologists. When an electrical impulse arrives at a synapse it does not cross the gap. However, the impulse causes some special chemicals, called *neurotransmitters*, to be emitted into the gap.

Neurotransmitters cross the gap, reach the dendrite, and start a process that changes the voltage at the dendrite. If the voltage change is positive, the neuron becomes more likely to produce an output pulse, while a negative voltage change makes the neuron less likely to produce an output pulse. Roughly speaking, the voltage changes caused by pulses arriving at all the dendrites of a neuron are added together, and if the sum becomes greater than a certain *threshold* value, the neuron fires a pulse. *This is decision making at a very low level.*

All this may be easier to understand by using an analogy of an official in a very polite bureaucracy. The official must decide whether to support a job applicant. The only way to deal with job applications in this bureaucracy is by sending (or not sending) *standard letters in favor of* a candidate. The official knows his subordinates who will send him letters. Each day he goes through all his mail and considers the letters he receives in favor of the candidate. If he receives a letter from a person whom he respects, he *adds one* to the sum of favorable letters. However, if he respects the sender a great deal, he *adds two* to the sum. For a letter from someone he does not respect, he *subtracts one* from the sum. At the end of the day, he looks at the sum. *If it is greater than ten, he sends a support letter to all of his superiors.*

In the analogy, the neuron (the official) decides whether to send a pulse (a letter) based on the sum of its input signals (the letters he received) after a *weight* (the numbers like $+1$, $+2$, or -1) is assigned to each input neuron (each sender). In the analogy, the *threshold* for sending a letter was ten. If a subordinate really liked a candidate, he might phone or send a letter to our official each day. Similarly, a neuron can transmit a *really strong signal* by sending out its maximum rate of pulses, perhaps a few hundred pulses per second. Thus, higher pulse rates from a neuron signal higher degrees of support for *whatever fact the neuron represents*.

One might ask what a neuron "means" when it fires at a high rate. If all pulses are more or less alike, and the rate of sending pulses only represents how *strongly* the neuron asserts what it is "saying," then how does the neuron pass on the *meaning* of its message? The answer is that *the meaning depends on which neuron is doing the signaling*. For example, suppose a neuron is connected directly to a receptor in the left ear that responds to sound at one hundred cycles per second. Pulses from this neuron "mean" that sound of the mentioned frequency is being detected in the left ear. The rate of pulses depends on how loud this sound is.

Now let's consider a less trivial example of the meaning carried by a neuron. Many experiments have studied how rats learn to find their way through a maze. In one study, a group of tiny wires was inserted into a rat's brain, and the activity of certain neurons in the *hippocampus* of the brain was observed as the rat passed through a maze that it had "memorized" previously. It was seen that a unique set of neurons fired for each particular place within the maze.[1] The *set of firing neurons* was a code for *where the rat was located*. The code for a given location was the same on different days when the rat visited the maze. In this example, the firing of a particular set of neurons appeared to "mean" that the rat was located in a particular place.

In the above example the neurons do not "know" the rat's location by magic. The neurons that signal the rat's location are connected, no doubt, by a complicated path through many other neurons, to neurons that carry sensory information. The sensory data brought into the rat's brain contain enough information to allow the rat to infer its location. The rat's brain determines its location by processing the sensory data.

Another example seems to show how animals recognize familiar objects even though the objects may be viewed from different directions. It appears that monkeys have stored within their brains a small set of "snapshots" showing a well-known object *viewed from different directions*. It was found that certain neurons within the *visual cortex* of a monkey's brain fired strongly when the familiar object was viewed at a particular angle.[2] The neurons fired weakly when the object was seen from other directions. For other objects, the neurons did not fire at all. It appears that the neurons respond strongly when the observed image matches the stored image of the object viewed at some particular angle. At other angles, the "match" is not as good, and the neurons respond weakly. For a very different object, no "match" is made, and the neurons do not respond.

Although a rat or monkey is not able to understand these abilities, the brains of these animals contain quite sophisticated features. By sight, the monkey is able to recognize many familiar objects. This implies that the monkey can identify these objects by detecting their features in the pattern of light falling on the retina of its eyes. This ability is called *pattern recognition*.

That a monkey has pattern recognition would not be surprising to people who work with *artificial neural networks*. These are sets of artificial neurons that are connected together to carry out practical tasks or to simulate the operation of parts of animal nervous systems. The simulations can be done on computers. Artificial neural networks are capable of recognizing patterns. In fact, their largest commercial application has been in problems involving pattern recognition.[3] Artificial neural networks, and presumably naturally occurring neural networks, are well suited for learning to distinguish different kinds of patterns. This would enable an animal, for example, to distinguish different kinds of objects or animals, and even to recognize specific animals, such as its litter-mates.

Identifying its own kind, its prey, or a predator is very important to an animal. Some animals seem to have certain *categories* "built-in" within their brains. Certain monkeys signal other members of their group when they spot leopards, eagles, or large snakes. The monkeys emit *different standard cries or calls for each of these kinds of animals*. This demonstrates that they distinguish the three kinds of animals by sight. These categories (the different kinds of animals) are not very abstract, and an animal may not be aware of having the categories, but they affect the animal's behavior.

An animal is probably not born with its brain "wired up" to identify all other kinds of animals. Most categories are learned and are embodied in neural networks that develop within the animal's brain. The specific features of the category are encoded within the connections of neurons within these neural networks. In humans, similar learning processes allow *concepts* to be learned. Humans give names to categories, and we can think about a category in an abstract way, without referring to a particular example of the item represented by the category.

LEARNING

The human brain contains something like 100 billion neurons. This huge number is about equal to the number of stars within our galaxy and the number of galaxies within the observable universe. The average neuron has from one thousand to ten thousand connections with other neurons. So, we would estimate that there are one hundred trillion (that's 10^{14}) or more connections within the brain. Could our genetic code, contained within our DNA, contain enough information to serve as a "wiring diagram" for making all the connections of neurons within our body? That is, *is the DNA tape long enough to specify where all the branches of the body's neurons should be connected?*

Human DNA contains about four billion base pairs. There are only four kinds of base pairs, so there is *no way* that the DNA can give a detailed specification of how to make one hundred trillion connections in the person's brain. In fact, Carl Sagan estimated that a person's brain contains a thousand times as much information as is contained in the person's DNA.[4] *The DNA cannot carry enough information for this colossal wiring job!*

The DNA, however, *does* carry a huge amount of information specifying the general organization of the brain. The DNA carries information for constructing mechanisms that enable neurons to make proper connections *with the help of experiences from the person's environment.* This picture is partly consistent with a conclusion of the English philosopher John Locke (1632–1704). He maintained that at birth a human's mind is empty, like a clean sheet of paper, and that knowledge is obtained by interacting with the world through the senses. Human infants and young animals resemble *learning machines*. They find it rewarding to explore and observe, developing skills and strength and acquiring knowledge about the world.

Knowledge can be brought to one's brain from the outside world by one's senses. We also can gain knowledge by the functioning of the brain itself, as when we reach a conclusion by reasoning. The process in which the brain acquires or incorporates new information is one form of *learning*. Through *memory*, some of what we learn is available to our conscious awareness at later times. In schools, teachers test this kind of learning by asking questions in examinations.

Another form of learning, acquiring *motor skills*, develops the ability to perform certain operations, such as riding a bicycle or playing a piano. Motor skills are tested by having the learner demonstrate the acquired skill. A person may not be sure that he or she possesses a learned motor skill. For example, I may be able to roller-skate, but I have not done it in many years, and I am not sure that I could skate now without having a few accidents. This kind of learning is not closely connected with our conscious awareness.

The processes involved in learning are under intense study, and much is being learned about learning. It is likely that learning involves making new connections between the brain's neurons or strengthening or weakening existing connections. Presumably, the learning process follows the laws of chemistry and physics. Learning sometimes involves detecting *associations* between different sensations or categories. The following example is probably not realistic, but it shows how learning might operate as an automatic process.

Suppose there is a neuron in someone's brain that signals, by pulsing rapidly, that a skunk has been detected. Suppose that some person can recognize a skunk from seeing them in movies, but she has never smelled the distinctive odor of a skunk. With a few experiences with a live skunk, however, she would literally form *a strong connection* between a neuron representing a particular pungent odor and the neuron which signals when a skunk is present. The odor-signaling neuron would become an important input to the skunk neuron. The skunk neuron would fire when the skunk odor was present, *even if a skunk was not seen*. As a result, the skunk odor would become associated with the category of a skunk in the person's brain, and she would think of a skunk whenever she smelled the skunk odor.

In the example, the learning *mechanism* might be the *strengthening of a previously weak connection* of an input synapse to the "skunk neuron." The connection might be strengthened automatically after several occasions on which skunks were observed and smelled at the same time. *The general process might involve strengthening of any connection between an input neuron and a following neuron when pulses are produced nearly simultaneously by the two neurons*. This would strengthen connections between neurons that represent occurrences that seem to be associated with each other.

MEMORY

Memory involves processes in which information is stored and can be retrieved at a later time. The *storage* of information in the memory process is a form of learning. The *retrieval* process (remembering) often involves a kind of search that is made easier by thinking of information that is connected with the information sought in memory. Of course, memory often gives us the information we desire immediately, and we are not aware of any searching process.

A certain kind of *artificial neural network* stores information so it can be retrieved based on *the nature of the information*.[5] This type of memory is called *content addressable memory*. It does not work the way computer memories work. Like human memory, it allows information to be retrieved with incomplete knowledge of related information.

Memories stored in human brains have been "played back" during brain surgery performed on people suffering from epilepsy. Since pain can be felt in the scalp, but not in the brain itself, this surgery was done on conscious patients using local anesthetics.

The surgeon touched an electrified probe to parts of the surface of the brain in order to find a region that seemed to the patient to "feel like" the damaged area where the epileptic seizures originated. The small problem area would then be removed, and the patient's condition usually improved. The surgery was done by a prominent Canadian neurosurgeon, Wilder Penfield. Some 8 percent of his patients reported vivid memories from their past when the probe touched their brain surfaces.

The memories would seem very realistic to the patient. The memories would include vision, hearing, and a sense of time and place. They seemed to intrude upon the consciousness of the patient, who would describe them to the surgeon. The memories were more vivid than usual memories, and the patients felt that they were *actual experiences from their own past*. The memories were evoked merely by using electric currents to stimulate certain areas of the brain.

One patient described such an evoked memory as follows: "I remember myself at the railroad station in Vanceburg, Kentucky; it is winter and the wind is blowing outside, and I am waiting for a train." Another patient reported hearing "a mother calling her little boy." The same experience occurred again, when the same position on the woman's brain was stimulated again. Upon stimulation of another location in her brain, she reported that she heard a man's voice and a woman's voice "down along the river somewhere," and she saw the river and felt that she had visited there when she was a child.

A similar experiment was done on an ape. The part of the brain involved in vision was stimulated. "The ape peered intently, as if at an object before him, made a swift catching motion of his right hand, and then examined, in apparent bewilderment, his empty fist."[6]

These results imply that memories are stored in a physical medium within the brain. Electrical currents can "replay" these memories. The experiments suggest that the normal retrieval of memories occurs by activating certain neural circuits in the brain. Perhaps the artificial electrical currents activate the neural circuit more strongly than usually happens naturally. This may account for the vividness of the evoked memories.

The results support the view that human *consciousness* is produced by electrical and chemical processes within the brain. The memory "replays" show that the experienced events do not need to actually occur for the experiences to be perceived as *real* by the person. *What is needed appears to be the proper activity within the brain*, which can be triggered by electrical currents. The behavior of the ape suggests similar conclusions concerning awareness in the ape.

ATTENTION

The processes of learning and memory appear to be passive and mechanical, while it seems that our brains are quite active in governing our experiences. The brain actively regulates our experiences by *attention*. Attention controls inputs from our senses and it controls the processing of information by the brain. As the American psychologist and philosopher William James (1842–1910) said, "My experience is what I agree to attend to."[7] *Attention transforms the brain from a sort of passive data-processing device into an active agent that determines its own activity.*

Our brains operate rather slowly. They can perform detailed processing of only a fraction of the abundant data arriving from the sense organs. Attention devotes the brain's finite resources to processing high-priority information. Attention can also prepare a person or animal to carry out a complicated operation more quickly or to maintain concentration on a task. It allows an idea, an observed object, or a plan of action to "fill the mind," even in the presence of the usual background sounds, images, and feelings. Thus, it must operate on the very broad regions of the brain that process sensory information, plan actions, and carry out other mental activities.

In a recent book, David LaBerge, a neuroscientist, discusses how attention may be implemented. He concludes that attention is a capability that is built into the brain's structure.[8] He thinks it is likely that a brain region called the *thalamus* amplifies or enhances selected information coming from other parts of the brain or the senses. The thalamus is connected by neurons to much of the brain, as it must be to play such a major role. Enough is known about the neural structures of the thalamus that LaBerge and his coworkers have been able to perform computer simulations of its operation. These simulations indicate that the circuitry of the thalamus could produce the enhancements of activity in selected parts of the brain that we call *attention*.

The thalamus appears to be an important means by which the "spotlight" of attention is focused on a certain part of the brain. According to LaBerge, the brain's *prefrontal cortex* is sometimes involved in *deciding where to aim the spotlight*. The thalamus also controls the operation of the brain during sleep and it may play a basic role in consciousness.[9]

APES, MONKEYS, MINDS, AND MIRRORS

"Common sense" leads us to believe that all kinds of apes are more closely related to each other than to humans. Until fairly recently, most scientists also believed this. This has changed as a result of studies done to compare the DNA sequences of apes and humans. Remember, DNA is the long ladderlike molecule that carries the "instructions" needed to form a new organism. The amount of agreement between the sequences of bases in the DNA of different species shows how closely the species are related to each other. The results can also be used to estimate the time interval since any two species shared a common ancestor.

The studies comparing the DNA of humans and the different species of apes show that humans and the African apes are very closely related and should be classified together. Humans, chimpanzees, and gorillas are more closely related *to each other* than they are to orangutans, the great apes of Asia.[10] Also, *most experts now believe that humans and chimpanzees are more closely related to each other than either is related to gorillas.*[11] The same studies have estimated that chimpanzees and humans had common ancestors approximately six million years ago.

Humans are one species, of course, but there are two species of chimpanzees. One is the common chimpanzee. I will call them *chimps*. The other is the pygmy chimpanzee or *bonobo*. If our ancestors acquired our special abilities by a long evolutionary process, we might expect that our closest relatives, the chimpanzees, would also display these abilities to some extent. Let's consider tool use, primitive forms of culture, self-recognition, and language in chimpanzees.

We are skilled in using tools. Depending on some subtle distinctions, tool use by animals may be evidence of high-level mental activity. Using tools is a way that an animal may demonstrate that *it understands its relationship to the surrounding world and can imagine ways that it can change this world to its own advantage.*

In their wild state in African forests, bonobos have not been observed to use tools, but chimps use a variety of simple tools. Some chimps get rainwater from holes in trees by dipping "sponges" made of chewed leaves or moss into the holes. Some chimps capture termites and ants by pushing twigs into the insects' colonies, jiggling the twigs to get the insects to attack the twigs, then pulling the twigs out with the insects attached.

Another impressive skill involves using a pair of stones to crack hard-shelled nuts. A large stone with a flat face serves as an anvil. A nut is placed on the anvil and a handheld rock is used as a hammer to crack the nut. Wild chimps usually learn to crack nuts when they are three to five years old, similar to the ages at which human children are able to crack nuts using similar tools.[12]

In another study, stones and nuts were set out for wild chimps near an observation station in order to observe the chimps' nut-cracking behavior. A stone was set out which might have served as an anvil except that its large flat surface normally was too far from being level. In three cases, different chimps learned to level the anvil stone by wedging another stone underneath it. The chimps had discovered a new method of cracking nuts using three stones: a hammer, an anvil, and a wedge! When the same stones and similar nuts were made available to human children, the youngest child who learned to use this method was six years and nine months old.[13]

In discovering the use of a stone wedge under an anvil, the chimps probably understood the problem of the slanted surface of the anvil and searched for a solution using nearby objects. They may have used *imagination* and *planning* to select plausible methods of solving the problem. Then, they tried various solutions and finally succeeded. Understanding a problem, and using imagination and planning are high-level mental activities that, in humans, are associated with consciousness.

Tool use has been observed in thirty-six different populations of wild chimps, and it is probably typical of the species.[14] Chimps in different areas differ widely in *which* tools they use. For example, chimps in West Africa crack nuts with rocks. Although nuts and rocks are both abundant in some more easterly regions, chimps in those areas do not crack nuts. At some locations, chimps ignore termites. In other regions they grab termites by hand, or trap them on small twigs, or dig for them. At most locations, no more than one method of capturing termites is used. This behavior, varying between different groups, has been described as a primitive form of *culture*. The observations support the view that cultural differences are not limited to humans, but also exist among different wild populations of chimps.

An experiment was done with chimps and with rhesus monkeys to see if they understand that *an observer can know something about an operation by seeing it.*[15] The animals were trained to try to find which of several upside-down cups contained food. The food had been placed in a cup while the cups were out of sight of the animal and one person (*the guesser*). The animal could see that the guesser was not able to see where the food was placed because the guesser had been out of the room or had placed a paper bag over his head. The animal could also see that a second human (*the knower*) had been in a position where he could see which cup covered the food.

Both humans pointed at a cup. The knower pointed at the correct cup and the guesser pointed at a different cup. The goal of the experiments was to see if the animal would "follow the advice" of the knower. The people often swapped roles, and the experiment was designed so that no simple rules could be used to identify the knower and the guesser, except that the knower had the opportunity to observe which cup contained the food, while the guesser did not.

Three of four chimpanzees learned to follow the advice of the knower, while the rhesus monkeys followed the advice of the knower or the guesser at random. The results suggested that three of the chimps (and none of the monkeys) understood that a person in a position to see an action has knowledge of what was done. More tests need to be done to see if there is a simpler explanation for the results.

Most animals are not able to recognize themselves in mirrors. Often an animal will react to its own mirror image as if it were seeing another animal. Some seem to learn that their images are not real animals and they tend to ignore mirrors. The animals that do not recognize themselves may not be aware of how their movements appear to an observer, or perhaps the idea that they could be seeing themselves "out there in the mirror" never comes to them.

Chimps and orangutans, however, are able to recognize themselves in mirrors. The case for chimps was established convincingly almost a quarter century ago and has been verified in other experiments.[16] At first, a chimp reacts to its images as if it were viewing another chimp. It soon learns that the image shows itself, and starts using the mirror as an aid in grooming parts of its body that it normally can't see. For example, it picks bits of food from between its teeth by observing its image in the mirror. It also observes its own behavior that is normally unobservable to itself, such as by making faces at the mirror.

The use of a mirror by a chimp appears to be evidence of a number of mental abilities. A chimp seems capable of imagining itself as having a body like that of another chimp. In this sense, the chimp seems to have a kind of self-concept. It also seems to be able to imagine how its motions appear to an observer, since it probably uses its motions to establish that the image in the mirror is of itself. With its ability to use tools, it is not surprising that a chimp can find a way to use the mirror to help itself with its grooming. The chimp is also able to deal with the fact that its mirror image is "turned around" and to overcome this complication in guiding its hand to clean its teeth.

The ability to recognize themselves in mirrors usually comes to chimps at about seven years of age, while it arrives in human children at a much younger age.[17] Gorillas, gibbons, and various kinds of monkeys failed to display this ability, although they have been tested extensively. Overall, the ability to recognize oneself in a mirror appears to be a test of very high intelligence in animals.

CHIMPANZEES AND LANGUAGE

Perhaps the ability to use language is the most uniquely human characteristic. Tests have been done to see how much ability chimpanzees have to use language. Wild chimpanzees communicate using simple gestures and a limited range of sounds. Straightforward attempts to teach captive chimpanzees to speak were failures: they could barely pronounce "mama," "papa," and "cup." Chimpanzees do not have the specialized vocal organs required for human speech.

Let's see how capable the chimpanzee brain is of supporting language even though chimps don't have our speaking abilities. Two American psychologists, Beatrice and Robert Gardner, produced a breakthrough when they began communicating with chimps by teaching them American Sign Language (ASL). Chimps have enough manual dexterity to form the ASL signs.

A female chimp named Washoe learned to use more than a hundred signs. Washoe and other chimps learned to communicate with each other and with their human trainers. (Washoe even tried, unsuccessfully, to communicate with the laboratory cat using ASL.) The chimps appeared to understand the meanings of the signs and they invented new two-sign combinations to describe things not covered by their limited vocabularies. For example, Washoe called a duck a "water bird." A watermelon became "candy drink" or "drink fruit" to a chimp named Lucy.

Chimps apparently are capable of generalizing the use of words to apply to new situations. For example, Washoe learned the sign "open" in connection with opening a door. She later used the same sign to describe opening a briefcase. This seems like an obvious use of "open" to English speakers, but Washoe had not been taught to use the word in this context. She had discovered a new context in which to use the word "open."

The first time Lucy ate a radish, she apparently found that it was too "hot," and from then on she described a radish as "cry hurt food." On one occasion when Jane, Lucy's (human) foster mother, left the laboratory, Lucy watched her leave and made the signs "Cry me. Me cry."[18] These examples suggest that a chimp's pain and emotions are "known" in the part of its brain that can communicate with other chimps or people. If a person reports having unpleasant feelings or emotions it is evidence that he or she has these feelings. Since a chimp has reported negative feelings and emotions to a human, doesn't it seem likely that the chimp also *has* these feelings?

The Gardners published their results on teaching sign language to chimps in 1969.[19] There has been a debate among scientists since that time concerning whether chimps really use language. Some later experiments have uncovered what seem to be additional language abilities in chimps.

The chimps Sherman and Austin learned to request things using a large keyboard with large symbols embossed on the keys. They later learned to announce that they were going to do certain things. For example, they would announce that they needed a certain key. Then they would search through a box of things to find a key to open a locked box. After special training, Sherman and Austin showed an ability to communicate information about absent objects. Furthermore, they were able to sort seventeen symbols into the two categories *food* and *tool*, using symbols for these general classes of things. To the experimenters, this demonstrated that the chimps associated the *properties* of named things with the *symbols* for the things.[20]

There is evidence that *bonobos* have at least as much language ability as *chimps*. A young bonobo named Kanzi learned to use a large electronic keyboard although *he* had not been trained for this. Kanzi learned to use the keyboard by watching the training of his mother. He soon surpassed his mother in using the symbols. Later, Kanzi learned to recognize spoken words. He could respond appropriately to novel requests, such as, "Kanzi, put some water on the carrots."

Kanzi was able to learn elementary rules of human grammar, satisfying five criteria that are supposed to be fundamental to grammar.[21] He also *invented* some rules of grammar. For example, he began using a keyboard symbol *before* a hand gesture in sentences in which he combined the use of the keyboard with gestures. I do not want to exaggerate Kanzi's abilities: the simplicity of the grammar used by Kanzi has been compared with that of a two-year-old child.

A chimp named Sheba was trained to associate the Arabic numerals 0, 1, 2, 3, 4, and 5 with the corresponding numbers of items in a group. In other words, she learned to count as many as five things. She was able to count a group of a new kind of object, say, flashlight batteries, on the first try. Thus, she seemed to appreciate that the *number* of things in a group does not depend upon what kinds of things are involved.

Without being trained to do this, Sheba developed a habit of separating any items in a group that were too close together and then pointing to each object once while counting them. She learned to find the total number of items in two groups. More remarkably, she learned to find the sum when she saw two cards labeled with Arabic numbers. *Effectively, she was able to add two small numbers to get a correct total.*

From the tool use, culture, self-recognition, language, and counting in chimpanzees, it appears that they are very capable animals. The common ancestor of chimpanzees and humans may have been about as intelligent as chimpanzees. It seems clear that the abilities of apes formed a solid foundation for the extensive further development of the same abilities in humans.

DINOSAURS, MAMMALS, AND PRIMATES

Let's discuss how and why humans are the dominant animals on Earth. First, I'll describe why *mammals* are the dominant large animals on Earth. We are mammals, of course. Mammals are warm-blooded and feed their offspring with the mother's milk. Warm-blooded animals can move quickly even when the surrounding air or water is cold. The ability of mammals to care for their offspring increases their survival relative to other animals. Mammals are quite intelligent animals.

Despite their advantages, mammals played a minor role on Earth during most of their history. The first mammals developed from mammal-like reptiles about two hundred million years ago. These mammals were mostly rat-sized or smaller and moved about at night. During the day they hid from the dinosaurs that controlled the world. In a vicious circle, the dominance of the reptiles probably kept mammals from evolving into larger forms that could compete with the reptiles.

About sixty-five million years ago the vicious circle ended. Earth was struck by an asteroid, a giant meteor more than six miles in diameter, traveling at twenty-five thousand miles per hour. The collision released a colossal amount of energy. The disaster affected the entire planet. Most larger animals, including all the dinosaurs, became extinct. Some small animals survived, including some mammals. With the dinosaurs gone, life became "a whole new ball game" for the mammals! There were no large predators or plant-eaters. Some mammals evolved into much larger animals. About ten million years after the extinction of the dinosaurs, *large mammals dominated Earth*. They have continued this dominance until now.

Humans are *primates*, the kind of mammal that also includes apes and monkeys. Most primates are adapted to life in the trees. Primates include lemurs, lorises, New World monkeys, Old World monkeys, apes, and humans. Unlike some tree-climbing mammals, primates did not develop claws. Instead, primates developed hands with five digits covered by soft pads. Primates use their hands and feet to grasp branches. They can also use their hands to pick fruit after spotting it by color vision.

Primates have forelimbs adapted to reaching for branches in a wide range of directions. They need to know how far away the branches are, too, and this is possible because of their *stereoscopic vision*. This requires that both eyes see the branches. Evolution accomplished this: both eyes moved to the front of the head. The faces of the primates became relatively flat, giving the eyes unobstructed, overlapping views in a wide range of directions. This was done by reducing the size of the snout, which is large and prominent in most mammals. A large snout is useful for animals that need a keen sense of smell, but this is not so important for animals living in trees.

Primates' brains deal with a three-dimensional world. To avoid fatal falls, they must control their movements consistently and precisely. Primate mothers take care of their young in trees, without a safety net. Most primates are social animals, with complex interactions within groups. These factors favor large brains, and primates *are* brainy animals. With their capable hands, strong arms with flexible movements, excellent eyes, and unsurpassed brains, primates were likely candidates to become manipulators of objects and, ultimately, manufacturers of tools. This probably had to wait, however, until some primates had adapted to living on the ground, so that their hands were free to handle objects.

HOMINIDS AND HUMANS

About fifteen million years ago, all of equatorial Africa was covered with forests, but this changed in the next few million years. In eastern Africa, some areas rose and formed high plateaus. Areas east of the highlands were left much drier, and the forests changed to patches of trees surrounded by grasslands. The *Rift Valley*, a major physical barrier, formed between the eastern and western regions. About eight million years ago the eastern region became still drier. It has been proposed that the common ancestors of humans and chimpanzees separated into two species at about this time, with chimpanzees in the western regions and our ancestors in the drier eastern regions.[22]

Perhaps the chimpanzees had an easier time, since they could carry on in the western forests much as they had in the past. The word *hominid* labels the relatively more human branch living to the east. Gradually, the hominids had to shift from living in forests to living in grasslands. Water and food for hominids were probably less abundant in the open country, making it necessary to travel long distances. For travel in open country, it is more efficient to have an erect posture and to travel by walking on two feet, rather than using the knuckles of the hands and the feet like the chimpanzees. The hominids eventually perfected erect walking.

Fossil *footprints* at Laetoli in northern Tanzania show two hominids walking near each other about 3.7 million years ago. There were also tracks of a saber-toothed tiger, giraffes, antelope, rhinoceroses, and elephants. Fossil bones of similar age from northern Ethiopia also show that hominids were well adapted for walking by that time. These discoveries show that upright walking was achieved before the hominid brain had become much larger than that of the chimpanzees and before hominids started making tools. The "free hands" that resulted from walking apparently did not lead immediately to tool making among human ancestors.

By 2.5 million years ago there were animals that have been classified as *Homo habilis*. They had larger brains than previous hominids. They also had different teeth, perhaps as an adaptation to a shift from a vegetarian diet to a diet that included meat. Because of these changes, I will call them *humans*, although they were far from modern in appearance or habits!

These early humans began making technical progress. They produced crude but sharp-edged stone tools by striking rocks together. The tools have been found near the bones of animals. On some two-million-year-old bones, scratch marks of the type made by stone tools were found, suggesting that tools were used to butcher animals and remove meat from bones.[23] In some cases, the scratch marks were made later than tooth marks of animals, indicating that our early ancestors may have been scavenging remains of carcasses that other animals had abandoned.

The *hand ax* was a very remarkable tool made by ancient humans. Hand axes are fairly flat stone objects with various oval outlines shaped like a melon seed or a teardrop, sometimes rounded and sometimes pointed at the narrow end. The wider end was often not sharpened, and was presumably held in the hand. The narrower or pointed end was sharpened on both sides, making the object useful for butchering animals, scraping hides, digging tubers, and cutting branches. It has even been suggested that hand axes were thrown into running herds of animals, like a sort of "killer Frisbee," with the goal of disabling an animal that could be killed by other means.[24]

Starting about 1.5 million years ago, large numbers of hand axes were produced *for over a million years*. Many hand axes are beautifully shaped, with accurate symmetry between the two sides. Such refined tools were made by individuals who *could visualize the desired tool* and *were skilled enough that they could shape the stone to match the goal*. These intelligent toolmakers probably belonged to the species *Homo erectus*. This species was possibly the immediate ancestor of our species, *Homo sapiens*. Fossil skulls show that the *H. erectus* brain size gradually increased from 850 to 1,100 cubic centimeters. This can be compared with an average of 1,350 cubic centimeters for a modern human and 400 to 500 cubic centimeters in chimpanzees and gorillas.[25]

The accomplishments of *Homo erectus* were not limited to hand axes. About seven hundred thousand years ago or earlier, these early humans left Africa and entered distant regions in what are now France, Indonesia, and China. They had learned to use fire to cook meat.[26] Presumably they also made clothing and shelters and used fire to stay warm in nontropical areas such as China and Europe.

WHY HUMANS DOMINATE THE EARTH

The ability to overcome physical limitations by manipulating objects is part of the reason why humans mastered Earth. In the open country, out of the trees, humans needed to defend themselves and drive predators away from carcasses. Although not well equipped with teeth and claws, our ancestors learned to throw stones, swing clubs, and jab with pointed sticks. Groups of humans formed armed and dangerous gangs.

Having flat faces lacking a snout, humans must have had difficulty removing meat from animal carcasses. Sharp stone tools were employed to cut through hide, remove meat, and break apart joints of animal bodies. Hammer stones may have been used to get at animal brains and marrow that were difficult for other animals to reach. Having evolved in the tropics, humans have little hair, but the use of fire, clothing, and shelters allowed them to invade the temperate zones.

Humans occupied *a unique and previously empty evolutionary niche*. We were the first animals to adopt a way of life based on manipulating our surroundings by making and using tools. Humans were destined to dominate Earth because this approach was *very general* and *open-ended*. It was very general in that humans could solve many kinds of problems by using or altering things in their surroundings. Solutions could be found to new problems that accompanied changes in the environment or migration into new regions. After about ten thousand years ago humans had expanded their range to include all the continents except Antarctica.

The niche occupied by humans was marvelously open-ended, since the possible gains from shaping the world to their needs were almost unlimited. Our ancestors started a process of invention that led from sharp-edged stones and pointed sticks to advanced tools and weapons, domesticated plants and animals, villages and cities, and all kinds of wonderful devices. The most remarkable feature of this way of life was that it allowed *continual progress*. New ideas and skills became part of the culture of following generations. Most animals have fairly stable habits throughout the lifetime of their species. Humans are different!

The improvement of the brain was a basic adaptation that helped humans fill this unique niche. To exploit the possibilities offered by culture and tool use, the brain needed to understand the shortcomings in the current way of life and to imagine ways to meet these needs. This required high *intelligence*. Our primate ancestors entered a way of life that depended on technical progress and culture. Once this process started, intelligence was useful in making rapid progress, so natural selection produced a larger and improved brain. Increased intelligence must have improved hunting, social, and communication skills.

The human way of life requires knowledge to be transmitted between generations. Children need to be trained in the ways of doing things. Even to produce a hand ax like those made a million years ago, a modern human adult needs several months of training and practice.[27] To transmit such knowledge, humans must have lived together in groups. Most likely this was nothing new. Our more apelike ancestors probably were social animals, like modern chimpanzees and gorillas. However, the transmission of knowledge did require something new: better means of communication, eventually including *speech*. Speech improved the communication between people. It increased their ability to work together, and presumably it improved social relationships.

Fossils display two weak hints of the use of speech. One is the shape of the brain, which affects the skull shape. A 1.8-million-year-old hominid fossil skull shows evidence of a shift from an apelike toward a humanlike shape in *Broca's area*, a part of the brain that controls speech.[28]

The second bit of fossil evidence comes from the shape of the base of the skull. It is associated with the position of the *larynx*. The larynx, the cavity which contains the vocal chords, has a different position in other mammals. The base of a 1.5-million-year-old *Homo erectus* skull indicates that its larynx was intermediate between the position in apes and in modern humans. A fully modern shape was only reached about three hundred thousand years ago, in early *Homo sapiens*.

The rather weak evidence mentioned above suggests that *H.*

erectus had some ability to speak, but that the full development of language ability occurred only in *Homo sapiens*. An improvement in speech or language ability may have been one of the reasons why *Homo sapiens* made such rapid technical progress, following much slower progress by *Homo erectus* during the million years of the hand-ax culture.

Humans surpassed other animals by entering a way of life based on *tool use, tool making, controlling their surroundings, living in groups, and culture*. This way of life preserved previously invented skills. New skills were included within a group's culture, increasing the technical level of the group. Increased intelligence improved the group's fitness by increasing the rate of invention of all sorts of new skills. Consequently, increased intelligence was produced within the members of the group by natural selection. The greater intelligence increased the rate of invention of new skills. *This situation resulted in an upward spiral of human mental ability and skills.*

The upward spiral of abilities resulted in advanced tool making, improved language skills, and greater intelligence. These developments led to cultural achievements such as art, religion, literature, and science. *Other species never entered this peculiarly open-ended niche and did not participate in an upward spiral of brain size and culture.* The apes remained speechless. They were capable of learning some basic elements of language, but never invented language. They continue to live a life that has changed little during the last few million years, with little tool use or culture. The upward spiral raised humans so far above the level of other animals that, when we are told that we are the closest relatives of the chimpanzees, our first reaction is to feel that this is not true.

Even in humans, however, all skills depend upon learning, memory, and other mental processes. These processes are based on interactions of neurons in an extremely complicated network within the brain and nervous system. Many human skills greatly exceed those of animals and seem completely removed from the capacities of nonliving matter. *Underlying these skills, however, are networks of neurons that operate by the same laws of physics and chemistry that apply to all matter.*

NOTES

1. Marcia Barinaga, "To Sleep, Perchance to . . . Learn? New Studies Say Yes," *Science* 265 (29 July 1994): 604.

2. David H. Freedman, "A Romance Blossoms Between Gray Matter and Silicon," *Science* 265 (12 August 1994): 890.

3. R. Beale and T. Jackson, *Neural Computing* (Bristol: Adam Hilger, 1990), p. 15.

4. Carl Sagan, *The Dragons of Eden* (New York: Ballantine Books, 1977), p. 26.

5. Beale and Jackson, *Neural Computing*, p. 191.

6. Sagan, *The Dragons of Eden*, p. 32.

7. William James, *Principles of Psychology*, 2 vols. (New York: Holt, 1890), 1:402.

8. David LaBerge, *Attentional Processing* (Cambridge: Harvard University Press, 1995), p. 218.

9. B. J. Baars and J. Newman, "A Neurobiological Interpretation of Global Workspace Theory," in *Consciousness in Philosophy and Cognitive Neuroscience*, ed. Antti Revonsuo and Matti Kamppinen (Hillsdale, N.J.: Lawrence Erlbaum Associates, 1997), pp. 211–26.

10. John C. Avise, *Molecular Markers, Natural History, and Evolution* (New York: Chapman and Hall, 1994), p. 331.

11. Richard W. Wrangham, Frans B. M. de Waal, and W. C. McGrew, "The Challenge of Biological Diversity," in *Chimpanzee Cultures*, ed. Richard W. Wrangham et al. (Cambridge: Harvard University Press, 1994), p. 5.

12. Tetsuro Matsuzawa, "Field Experiments on Use of Stone Tools by Chimpanzees in the Wild," in *Chimpanzee Cultures*, p. 356.

13. Ibid., 361–64.

14. W. C. McGrew, "The Material of Culture," in *Chimpanzee Cultures*, p. 28.

15. Daniel J. Povinelli, "What Chimpanzees (Might) Know about the Mind," in *Chimpanzee Cultures*, p. 289.

16. Gordon G. Gallup Jr., "Chimpanzees: Self-Recognition," *Science* 167 (2 January 1970): 86.

17. Povinelli, "What Chimpanzees (Might) Know about the Mind," p. 293.

18. Sagan, *The Dragons of Eden*, pp. 117, 118.

19. R. A. Gardner and B. T. Gardner, "Teaching Sign Language to a Chimpanzee," *Science* 165 (15 August 1969): 66.

20. Duane M. Rumbaugh, E. Sue Savage-Rumbaugh, and Rose A. Sevcik, "Behavioral Roots of Language," in *Chimpanzee Cultures*, p. 324.

21. Patricia Marks Greenfield and E. Sue Savage-Rumbaugh, "Grammatical Combination in *Pan Paniscus*: Process of Learning and Invention in the Evolution and Development of Language," *"Language" and Intelligence in Monkeys and Apes*, ed. Sue Taylor Parker and Kathleen Rita Gibson (Cambridge: Cambridge University Press, 1990), p. 541.

22. Yves Coppens, "East Side Story: The Origin of Humankind," *Scientific American* 270 (May 1994): 88.

23. Richard Leakey, *The Origin of Mankind* (New York: BasicBooks, 1994), p. 70.

24. William H. Calvin, *The Ascent of Mind* (New York: Bantam Books, 1990), p. 181.

25. Leakey, *The Origin of Mankind*, pp. 47, 82.

26. Brian M. Fagan, *People of the Earth*, 7th ed. (New York: HarperCollins, 1992), p. 131

27. Leakey, *The Origin of Mankind*, pp. 39, 40.

28. Fagan, *People of the Earth*, p. 114.

9

CONSCIOUSNESS, THE SOUL, AND THE BRAIN

WHATEVER ELSE IT ACCOMPLISHED, THE adoption of Christianity by the Roman Empire and later by almost all of Europe locked that part of the world into a deep faith in the supernatural. This set of beliefs differs greatly from the predictable orderliness of the scientific worldview.

Some people are able to accept both the Christian and the scientific worldviews without being troubled by the differences between them. Some others are deeply impressed by the order and unity of the scientific worldview, but are not able to believe in the world of the supernatural. An arbitrary and unpredictable *realm of the spirits* is contrary to their idea of reality. They argue that supernatural beliefs vary from one creed to another and that no religion can give convincing evidence for its supernatural beliefs. They conclude that no set of religious beliefs is worthy of steadfast devotion.

Supernatural beliefs are losing their usefulness as science explains more and more about the world in terms of natural and orderly processes. In this and the following chapters I will argue against the belief in a supernatural entity that is supposedly intimately connected with our existence. The criticized idea is that a person consists of a supernatural soul, besides a material body.

Philosophers call such a belief in two parts of human nature *dualism.*

Dualism is intimately connected with our ideas about ourselves. I will argue that no part of a person, including consciousness, is supernatural. The argument will be that believing in the existence of a human soul does not help to explain anything about ourselves. Since so much of the reason for believing in a soul arises from the apparent mystery of consciousness, the main issue will be whether consciousness may be produced by purely natural processes.

LIFE WITHOUT THE SUPERNATURAL SOUL

If we shift our view back to the time of the origin of Christianity, we see that many changes have occurred in the set of things that are considered to be supernatural. Now, we rarely believe that certain kinds of disability or illness are due to possession by an evil spirit. At the time of Jesus, there were still some reports of evil spirits flitting around the landscape of Galilee, but they now appear to be an endangered species.

Many of the ancient ideas about the soul are no longer held. The soul is no longer thought to be associated with one's blood or breath. Most people do not believe in a soul that leaves the body during sleep or trances. The soul is no longer thought to be involved in such life processes as metabolism, by which food is used to provide materials and energy to the body. The motions of an animal's body are fairly well understood without invoking the idea of a soul. On the other hand, the idea of a soul survives and is used to explain those human abilities that are still regarded as mysterious by some religions.

One ancient reason for believing in the soul was that it could explain the differences between a live and a dead person. In chapter 7, we considered the origin and evolution of life without mentioning a soul or any other supernatural basis for life. Modern biology does not suggest that there is anything supernatural about the origin or existence of life. Biology recognizes that modern life is part of a continuous process reaching back billions of years. Our ancestors include one-celled organisms who lived in the oceans when Earth's continents were still barren. We share in the awesome process of the development of life on Earth.

We saw that the processes of life are special in that they are necessary for life's continuation, but, at the atomic level, these processes are the same as those that occur in nonliving matter. The chemistry of living and nonliving matter is the same. *The differences between living and nonliving matter are found in the elaborate and intricately organized structure of living matter.* Living matter is the highly refined product of billions of years of testing and modification of self-reproducing structures. Living matter is a completely natural and beautifully organized product of Earth's unusually favorable environment.

Ancient peoples did not have this knowledge of the nature of life. They made the mistake of believing that the difference between living and nonliving matter is the presence or absence of a *thing*, the life soul. From a more modern point of view, the difference between living and nonliving matter is not caused by the presence or absence of some entity. Living matter has *an inherited organization and functioning* that comes from the long history of life on Earth. During the growth of a living being, nonliving matter enters the body and becomes living matter. Death is the ending of the functioning of the life processes, leading to the loss of the special organization of the matter. The formerly living matter becomes nonliving matter.

We can see why ancient people believed in a life soul, but we can also understand why many modern people do not believe in a life soul. For many, this idea has outlived its usefulness. Scientific progress has made the idea obsolete. In the following sections I will argue that the traditional Christian soul is also becoming obsolete. Scientific explanations are being found for some of the traditional functions of the soul. *These explanations do not involve a soul at all.* Other traditional soul functions are not understood now, but many scientists expect that they will eventually be explained by science. *Still other traditional soul functions simply do not exist*, in the opinion of many philosophers and scientists.

MENTAL MYSTERIES, THE SOUL, AND THE BRAIN

To ancient people and to many modern people it has been impossible to imagine that our mental processes, such as remembering and imagining, arise from interactions between atoms within our bodies. For many, the inability to imagine this supports the belief that there is a supernatural side of our existence. Among our not so distant ancestors, the known physical processes were the functioning of such things as falling objects, ropes and pulleys, hooks and push rods, and, at the limits of imagination, the movements of planets in orbit. It was natural to feel that it would be impossible for mechanical systems to produce something as immaterial and mysterious as thinking and consciousness.

Not being able to imagine something does not prove that the thing could not exist. For example, a person living during the Middle Ages would not have been able to imagine that various materials such as sand, metal ores, and an almost unknown material called *petroleum* could be processed and made into a television set or a computer that could help control the motion of an airplane. It would, of course, have been a mistake for the person to conclude that such things are impossible.

At first sight, the electronic circuits inside a television set or a computer do not appear to be able to achieve what actually occurs within these devices. Similarly, in butchering or dissecting a dead animal, it is not evident that this *thing* could see things and decide how to move in various complicated situations. In spite of appearances, however, electronic devices and live animals can function by themselves: they do not work by magic or with the help of supernatural entities. After considering these examples, we should not be too skeptical of what can be achieved by the jellylike substance inside the rigid container between our ears.

To people of, say, the first century C.E., it was difficult to imagine that human mental abilities could be attributed to the brain. A few centuries earlier, Aristotle thought that the brain's function is *to cool the blood.* For many inhabitants of the Roman Empire, it seemed sensible to believe that at least some mental abilities are supernatural, and, more specifically, that they are imparted by the soul. This became part of Christian dogma and was held by most Europeans after their empire, tribes, and nations converted to Christianity. The belief in a soul was also transported to the European colonies, of course, where this official doctrine from the later days of the Roman Empire is still commonly believed.

Although Christianity taught people to believe that an immortal soul is the source of the human ability to reason and the human power of free will, the way in which the soul gives humans these powers was not explained and was treated as a mystery. Thus, the soul was not really an *explanation* of the existence of the mental abilities, but rather it was a *name* for one of the *"mysteries of the faith."* Belief in the soul tended to discourage scientific thinking about the mind. Even now there is still something basically mysterious about "the mind."

For a long time, belief in the soul was part of the mainstream of philosophy as well as religion. As described in chapter 2, René Descartes was very important in establishing a modern philosophical basis for the belief that a human being lives a dual existence involving a spiritual soul and a body. Descartes believed that the body and soul must interact with each other, and he thought that this interaction occurs through the pea-sized *pineal gland* within the brain. The gland's actual function is to produce and secrete the hormone *melatonin,* which appears to be involved in biological rhythms and reproductive behavior in animals and humans.

Although it no longer appears plausible that the pineal gland has a role in communicating with the soul, other residues of Descartes's thinking still influence our worldview. The soul belief is so basic in our culture that, through ordinary communications, most of us come to believe that a network of neurons cannot, by itself, generate our thoughts and our awareness of the world.

People are often willing to accept the idea that consciousness is produced by a *completely mysterious activity* of the soul, but people are usually *not* willing to accept an explanation of consciousness involving only the brain without receiving a convincing description of the proposed mechanism. This *biased approach to the issue* probably comes from the ancient idea that consciousness is, by its nature, occult or mysterious. In this view, a *natural* explanation of consciousness *could not possibly be adequate*.

Ideas about the mind were strongly affected by the publication, in 1949, of *The Concept of Mind*, by the British philosopher Gilbert Ryle (1900–1976).[1] In this book, Ryle attacked the "dogma of the Ghost in the Machine." This dogma maintains that there is an immaterial entity, the mind, that is similar to the soul as we have described it. Ryle considers it a basic philosophical mistake to put the mind in the same category as the body. The mistake is to regard the mind as a kind of immaterial *thing* or *substance*. A majority of present-day philosophers would agree with Ryle on this.[2]

A definition of mind that avoids the mistake mentioned above is that *the mind is a label for one's ability to have mental processes occur within one's brain*. In this definition, the mind is regarded as merely a name for a bundle of capabilities, such as the capacity of a functioning brain to carry on such processes as feeling, thinking, remembering, deciding, initiating actions, imagining, planning, and so forth. The word "mind" describes a portion of the abilities that one has as a living human being.

From 1913 until about 1965, there was almost a double taboo on studies of the mind. During this period, most American psychologists accepted the views of *behaviorism*. Behaviorism maintains that psychology is the use of strict experimental procedures to study the *observable* behavior of individuals interacting with their environment. In this picture, a person or animal is often regarded as a "black box" that receives stimuli from the environment and produces behavior in response. The internal functioning of the black box is often ignored. Psychologists of the behaviorist school tend to avoid speaking of the mind as such, because the word suggests the existence of an unscientific entity like the soul, which could not be detected by experiments.

During the same period when behaviorism dominated psychology, our everyday culture was permeated with the Christian belief that the really interesting mental abilities are supernatural and mysterious. This implied that a scientific study of consciousness, for example, was misguided, hopeless, and irreverent.

In recent decades, barriers to the scientific study of the mind are being removed. Attempts are being made to understand the information processing and mental processes that occur within the brain. In current scientific literature, mental processes are often referred to as *cognitive functions*. Since the so-called cognitive revolution in psychology, psychologists are renewing their study of the normally invisible mental processes that underlie our behavior.[3] By the late 1970s, psychologists started to use the word "consciousness" again.[4]

In recent years, some books have made especially valuable contributions to the public's understanding of the mind's nature. *Consciousness Explained*, by Daniel Dennett, is an important and popular book of this type.[5] Dennett, an outstanding American philosopher, discusses how consciousness may arise. Another valuable book on this subject is *The Astonishing Hypothesis: The Scientific Search for the Soul*, by Francis Crick.[6] Crick, introduced in chapter 7 as one of the discoverers of the structure of DNA, describes how scientific studies of the brain are helping us understand mental processes. Gerald M. Edelman, an American brain scientist and Nobel prizewinner in physiology in 1972, has written an interesting book named *Bright Air, Brilliant Fire*, describing his theory of the mind.[7] A very interesting view of brain science from a psychological perspective can be found in Bernard J. Baars's *In the Theater of Consciousness: The Workspace of the Mind*.[8] It uses the theater as a metaphor in describing the activities of the conscious brain. The theater metaphor resembles the picture presented by the "toy model" in the next chapter.

THE BRAIN'S ROLE IN MENTAL PROCESSES

The biggest problem in the scientific understanding of mental processes occurring within the brain is *the severe difficulty of the subject*, rather than the unfavorable attitudes of many individuals. One of the most outstanding features of the brain is its enormous complexity. The detailed understanding of a mental process might require tracing signals in individual neurons within the brain, where there are something like one hundred billion neurons. The neurons themselves are extremely tiny, and there are few situations that allow electrical probes to be inserted into the brain of a living human being. In spite of these difficulties, brain scientists are making substantial progress using information obtained from a variety of sources.

Let's discuss some of the evidence that shows that the brain is involved in all of our mental activities. We all frequently have our brain partially shut down when we sleep. During sleep we are much less sensitive to the responses of our senses. While asleep we sometimes enter an altered form of consciousness that is not closely linked to what is actually happening outside the body. *Dreaming* seems to be a spontaneous form of consciousness that sometimes arises in a free-running brain in a situation in which inputs from the outside world do not guide the brain's operation. *Dreaming suggests that the brain is a sort of consciousness-generating organ that can go into another mode of operation in which it produces consciousness spontaneously while cut off from signals from the outside world.*

In chapter 8, we saw that electrical stimulation of the surface of the brain can trigger conscious experiences that appear to be stored as memories within the brain. Similar stimulation of brain tissue by electrical probes has produced *very strong emotions* in subjects having brain surgery with only local anesthesia. During excitation of a point on the *temporal lobes*, a patient experienced a very pure sense of dread but couldn't explain what it was that he dreaded. By touching various locations it was possible to stimulate fear, guilt, loneliness, and disgust, as well as "an ecstatic feeling that all problems are soluble."[9] These effects demonstrate that memories and emotions are generated by the brain.

We also discussed, previously, how neurons may implement the process of learning by making new connections with other neurons. We saw that pattern recognition in monkeys seemed to involve a number of stored images of certain objects seen from different points of view. A particular neuron was observed to fire when a monkey saw an object from a specific direction. We discussed a theory of how attention may be implemented by the thalamus, a part of the brain. Thus, memory and consciousness, emotions, learning, pattern recognition, and attention all seem to be connected with brain activity.

Many mental functions have been connected with the brain through observing that those functions can be disabled by injury to parts of the brain. People with brain damage from strokes, gunshot wounds, and concussions have suffered loss of consciousness; learning disorders; loss of various kinds of memory; and the inability to speak, concentrate, detect movements of objects, or recognize faces. Others have suffered lack of visual depth perception, partial or full blindness, loss of the ability to smell, and other problems.

We all know that various drugs and chemicals can affect mental functions by affecting the central nervous system. Alcohol can slow one's ability to react, weaken one's judgment, cause the surroundings to appear to rotate, and disrupt one's sense of balance. LSD can produce hallucinations. Various anesthetics can cause one to lose the sense of feeling or even to lose consciousness. Caffeine can affect one's alertness and ability to sleep. Other drugs can produce feelings of well-being, improve one's ability to relax, and relieve depression. Many mental illnesses can be effectively treated by various *psychoactive drugs*. These drugs that affect the mind include alcohol and opium, known even in ancient times, and modern tranquilizers and antidepressants. Since the 1950s, drugs have been found that are beneficial in treating such mental disorders as depression and schizophrenia.

Since drugs can affect one's emotions and consciousness, it is not surprising that chemical processes within the brain can produce natural drugs with similar effects. This was noted by Antonio R. Damasio, a prominent American neurologist. "The well-known relationship between chemistry and feeling has prepared scientists and the public for the discovery that the organism produces chemicals that can have similar effect. The idea that endorphins are the brain's own morphine and can easily change how we feel about ourselves, about pain, and about the world is now well accepted. So is the idea that the neurotransmitters dopamine, norepinephrine, and serotonin, as well as peptide neuromodulators, can have similar effects."[10]

Clearly the brain is deeply involved in mental processes. For us to be mentally normal, our brains must be properly organized and nourished by a bloodstream that is free of harmful drugs or chemicals. The normal brain must not be damaged by strokes, diseases, concussions, or penetration by objects. So, *mental activities require a functioning brain.*

CONSCIOUSNESS AND THE BRAIN

Unlike the pagans and early Christians who planted the firm belief in the existence of the soul within our culture, *we* know that the brain is involved in all mental processes. Let's consider the possibility that *the brain is all that is needed for consciousness.* That is, let's consider the idea that consciousness is not dependent on a soul, but is a natural result of the functioning of the brain. For our purposes, we can define consciousness as *what it is like to be a person who is awake or dreaming and has a normally functioning brain.* Perhaps this definition should include some animals, but we are not certain that they are conscious.

By our definition, consciousness is interrupted by dreamless sleep, and it returns when we awaken or have a dream. By almost anyone's definition, consciousness leaves when a person is under general anesthetic during surgery. The fact that consciousness can be halted and restarted is evidence that it is due to the operation of a *process*, rather than the presence of *a spiritual entity*. This is consistent with the view that consciousness arises from a dynamic process within the brain, rather than from the presumably continuous indwelling of a soul. The idea that consciousness is the result of a process is not new. Gerald Edelman has noted that the American philosopher and psychologist William James (1842–1910) proposed this idea in an essay entitled "Does Consciousness Exist?"[11]

The mental processes in the brain are enormously complicated. Human vision is an outstanding example. The processing of visual information starts within the retina of the eye, in networks of interacting neurons. Bundles of neurons carry information from the eyes to the brain. In Francis Crick's book *The Astonishing Hypothesis*, there is a diagram showing the connections between the different areas of the brain that are involved with vision.[12] There are forty-seven areas identified in this diagram, each with different functions. There are millions of neurons in each of these areas, and each neuron may connect to thousands of other neurons.

The brain's processing of visual images is broken down into separate tasks such as detecting brightness variations and color in two dimensions and identifying visual features such as straight lines and edges of objects. Separate neurons deal with lines oriented at different angles and located at different positions within the field of view. Objects are identified by pattern recognition and distances to the objects are estimated using the small differences of directions to the object arising from the separation of the eyes. Pattern recognition involves great amounts of stored information. We probably identify various individuals we know by storing information about each person's appearance when viewed from various directions.

Motions of visual patterns are analyzed in detail. One brain region deals with moving objects that appear to be growing in size. This has survival value in drawing attention to approaching objects and animals.

As the processing of a visual pattern proceeds, it gradually involves more sophisticated discriminations. The earliest stages of processing handle such basic information as the brightness of light in different directions and the colors associated with this light. Further on, edges of objects are identified and identified edges are then passed on to pattern recognition functions. Eventually such patterns as the face of a known person are identified.

After a face is identified, it can be analyzed for evidence of emotions. Visual information that a face displaying a certain emotion has been identified may then be combined with information from other sources. Short-term memory may furnish the preceding images of this person and the words he or she spoke. The spoken message initiates activities dealing with recognizing and understanding the words and realizing the meaning and implications of the entire message.

Altogether, the activities described above produce an awareness that a known person is communicating some particular message. While one is reacting to the message, other parts of the brain may be monitoring the setting, such as the location and surroundings. Meanwhile, emotions, drives, and the level of alertness, generated by various parts of the brain, are affecting the person's overall impression of the situation.

Clearly, the processes of consciousness are complicated. *Enormous numbers of neurons are needed to carry out all the functions mentioned above.* Memory must involve huge numbers of connections between neurons in order to store all that we remember.

Note that we are reaching conclusions that are the opposite of some of Plato's. He believed that the basis of mental activity is an immaterial entity with no parts. Our discussion implies that this role is played by a material entity, the brain, with a fantastically large number of parts. Modern knowledge of the brain leads to the conclusion that *complexity, involving enormous numbers of components, is one of the necessary conditions for human consciousness.* In chapter 2 it was mentioned that Plato argued that a spiritual entity with no parts, such as the soul, should be immortal, since it could not fall apart. Since a conscious entity involves so many parts, and following Plato's reasoning, it seems that conscious entities are not likely to be immortal.

USING A BRAIN TO REPRESENT REALITY

In this and the following sections, I will argue that consciousness was a natural result of the evolution of animals. This discussion is fairly speculative, but it will expand our ability to understand how consciousness may arise by natural processes. Let's consider why animals developed nervous systems and brains that led to the emergence of consciousness.

Animals need to find food and avoid being eaten. Animals evolved sensory organs and nervous systems to select and control their movements in ways that helped them survive. Complicated nervous systems developed in many animals. For example, even a simple scavenging animal that is guided to food by smell may have very many neurons. Sense organs and neurons can lead the animal toward the smell and detect when the animal arrives at a piece of food. The animal can then stop and move jaw or mouth muscles in order to eat.

Even such a simple animal may need *connections between neurons* to help find and eat food. In an advanced animal, moving and eating would each require many nerve connections in order to coordinate the contraction or relaxation of different muscles involved in the two processes. Animals with complicated nervous systems need extremely large numbers of connections between neurons. Perhaps the *brain* arose as a central "switchboard" in which all of these connections could be made more efficiently in order to minimize the total length of the connections. *Animal brains are large masses of connected neurons that guide the animal's movements and control many bodily functions.*

When animals acquired vision, it made their nervous systems more complicated. At first there were only a few light-sensitive cells that responded to light coming from a wide range of directions. Vision rapidly became more refined, allowing animals to detect detailed two-dimensional views of the world. Enormous numbers of neurons were devoted to processing the abundant information made available by the eyes. With detailed vision it became possible for animals to use pattern-recognition processes to recognize other animals by sight. Animals became able to distinguish predators or enemies from harmless animals or food animals. Animals also became able to learn and remember the behavior of other animals so that they could respond appropriately when they spotted those animals.

Valuable studies of vision have been done with the help of fully conscious patients awaiting brain surgery. The patients, whose skulls had been opened, required only local anesthetics to avoid pain. Small lights were turned on in various locations viewed by the patient. Electrical probes detected the activity of nerves in the patient's brain responding to the light. This work showed that there is a detailed correspondence between directions of the light and locations of affected neurons within the brain's *visual cortex*. Drawings have been made that show that light observed from different directions is *mapped* in a very orderly way onto the patient's visual cortex.

More recently, it has been found in work with animals that there *are a number of different "maps"* of the visual world at different locations within the animals' brains. The rat has six such maps, the owl monkey has at least eight, and the cat has at least thirteen maps.[13] The various maps are associated with the processing of *different kinds of information* about the world. For example, separate maps are involved with detecting color, edges of objects, sizes of objects, orientation of objects, and movement of objects.

When we look at a scene such as the street outside an office window, we perceive images of various objects. There may be sidewalks, people, moving and stationary vehicles of various colors, and buildings, plants, and a clear or cloudy sky. We usually do not have to pay attention in order to recognize these different kinds of objects. We can identify the kind of object "immediately" upon "seeing" the object. We are not aware of the operation of all the neurons in the eye and the brain that are analyzing brightness variations between neighboring light-detecting cells in the retina of the eye. We are not aware of all that is happening within the brain that enable edges of objects to be traced.

In viewing a scene we are not aware of the operation of the pattern detectors that identify the objects as cars, people, or trees. We are not aware that our brain takes the image from the retina and sends it to various areas of the brain to be processed. After complicated processing of different characteristics of the objects by several parts of the brain, a person becomes aware of the entire scene, with the right colors "painted" onto the appropriate objects. The objects are usually *immediately recognized* as cars, people, and so on, with some objects appearing stationary and some moving. We are also immediately aware of the *distances of nearby objects*, which the brain finds by analyzing the effects of the slight differences of perspective from the left and the right eye.

We are aware of only some of the processes that are going on when we observe a scene. Many of the other processes involved in dealing with the visual data are unconscious or *subconscious processes*. A computer expert might say, "These processes are transparent to the user." At the conscious level, we get the valuable results of all these processes. When asked why a scene looks a certain way, we may reply, "Because that's the way it really is." We may believe that the scene constructed by the brain is an unaltered presentation of reality. The fact that we can believe this shows that the brain does its job very well. We *can* usually believe our senses. It is so unusual that our perceptions are found to be wrong that we may be seriously disturbed if it happens.

What the brain produces, when we view a scene, is not simply a view of reality. It is *a specially constructed version of reality* that serves our need to survive. We know that the brain can construct representations of reality even in the absence of sensory data. This frequently happens to us when we dream and when we imagine or remember images or events. When we see a scene, the patterns of light that fell on the retinas of our eyes have been processed, taken apart, and effectively put back together to form a scene that contains information that is likely to be of survival value to us.

When we see a scene, pigs, panthers, and peaches are identified in our awareness, along with information about whether they are moving and how far they are from us. As a product of biological evolution, *the brain developed as a system for gathering, pro-*

cessing, storing, and retrieving information that enables the owner to select and carry out those movements that give the animal or human the best chance of surviving. The brain also carries out essential "housekeeping" duties, such as regulating breathing.

As a basis for choosing actions, the brain gathers information and acquires knowledge. Young animals need to learn to do many things in order to survive. Just walking on uneven ground requires a steady stream of information about rocks, holes, and fallen branches. Animals need to know where to find water and food. They need to recognize other kinds of animals and know their habits. It is useful for social animals to learn to identify their companions and to know each individual's status and temperament. Such mental processes as learning, memory, and pattern recognition help the animal meet these needs. In chapter 8, there was a brief discussion of how making connections between neurons in neural networks enables animals to carry out these processes.

Because there is much orderliness in the world, the brain can detect patterns in the properties of objects and the courses of events. These patterns are the basis for *concepts* that are formed by making neural connections within the brain. By learning about the world through many experiences, the human or animal brain builds up a *representation* of the world and how it works. For example, pattern detectors may learn to recognize skunks. The person or animal may have learned that skunks can produce a strong-smelling substance and that skunks can use this substance to defend themselves successfully even though they can't outrun most larger animals. After learning these properties of skunks, the person or animal will be able to identify skunks, predict their behavior, and avoid making certain mistakes.

We know that in humans the ability of the brain to represent the world is so strong that we can *imagine* how events may proceed in many situations. Our brains, or should I say *we*, can predict how certain people or things will act based on our previous experiences. However, not all of the brain's power to represent reality is due to learning from experiences.

Our remarkable ability to estimate distances of nearby objects depends on a built-in image-processing system that converts two slightly offset two-dimensional views of a scene into a three-dimensional image. Presumably our DNA contains instructions for setting up this special system. In other words, we *inherit* this brain structure that helps us make three-dimensional representations of our surroundings. *In fact, much of the remarkable ability one has to construct a realistic model of the world within one's brain is probably inherited.* The structure of the brain has evolved to allow a useful representation of the world to be made with the assistance of our senses and our experiences.

NOTES

1. Gilbert Ryle, *The Concept of Mind* (New York: Barnes and Noble, 1949), p. 15.

2. Daniel C. Dennett, *Consciousness Explained* (Boston: Little, Brown & Co., 1991), pp. 33, 37.

3. Bernard J. Baars, *The Cognitive Revolution in Psychology* (New York: Guilford Press, 1986), chapters 1 and 4–7.

4. Ibid., p. 175.

5. Dennett, *Consciousness Explained*.

6. Francis Crick, *The Astonishing Hypothesis: The Scientific Search for the Soul* (New York: Simon & Schuster, 1995).

7. Gerald M. Edelman, *Bright Air, Brilliant Fire: On the Matter of the Mind* (New York: BasicBooks, 1992).

8. Bernard J. Baars, *In the Theater of Consciousness: The Workspace of the Mind* (New York: Oxford University Press, 1997).

9. Peter W. Nathan, *The Oxford Companion to the Mind*, ed. Richard L. Gregory (Oxford: Oxford University Press, 1987), p. 527.

10. Antonio R. Damasio, *Descartes' Error* (New York: Grosset/Putnam, 1994), p. 160.

11. Edelman, *Bright Air, Brilliant Fire*, p. 37, refers to William James, "Does Consciousness Exist?" in *The Writings of William James*, ed. J. J. McDermott (Chicago: University of Chicago Press, 1977), pp. 169–83.

12. Francis Crick, *The Astonishing Hypothesis*, p. 156.

13. Peter Alexander, "Localization of Brain Function and Cortical Maps," in *The Oxford Companion to the Mind*, pp. 436–38.

10

CONSCIOUSNESS AND
MODELS OF THE MIND

C ONSCIOUSNESS IS ONE'S MOST PRECIOUS possession. Although it is
the essence of one's existence as a person, it seems mysterious because it is poorly understood. The mystery of consciousness may lead a person to think that it has a supernatural origin. In
traditional Christian teaching, one's consciousness comes from a
supernatural soul. This and the following chapter will argue against
that belief.

The idea that consciousness is supernatural is so deeply embedded within our culture that we are strongly prejudiced against
accepting *natural* explanations of it. Consciousness seems to be outside the domain of natural processes. To some ancient people, it was
inconceivable that physical systems could produce things as immaterial as thinking and consciousness. Even today, many people can't
imagine that our mental processes, such as remembering and imagining, could possibly arise from interactions between atoms within
our bodies. This *failure of imagination* leads many of us to believe
that our existence is tied to supernatural entities.

Unlike the ancient thinkers who founded our culture's strong
faith in the soul, we know that the brain is involved in all mental
processes. This was discussed in the previous chapter. Now, let's

consider the possibility that *the brain is all that is needed for consciousness*. That is, consider the idea that consciousness does not depend on a supernatural soul, but is a natural product of the working brain.

A simple "toy model" of consciousness will make it easier for us to imagine how the brain produces consciousness. The model is an educational toy. By playing with it we can appreciate how consciousness arises in a brain that has the proper structure. *This model explains how some of the most mysterious properties of consciousness may arise within our brains*. Let's highlight some properties of consciousness that the model can help us understand.

In discussing vision in the previous chapter, we saw that a large portion of the brain carries on subconscious activities. For example, the processes by which visual objects are constructed are subconscious. We become aware of a fully constructed object, but not the processes that detected and identified this object in the pattern of light falling on the retinas of our eyes. One way that we may understand consciousness is by learning why some brain activities are conscious and others are not.

Consider some properties of conscious activities. Conscious mental activities enter what has been called "working memory,"[1] where they are *remembered* for a short time. We can *speak or write* about these experiences, they are available to our process of introspection, and they may form long-term memories.

After a conscious experience, we often go on to related thoughts. For example, you may see a child riding a bicycle for the first time and this may lead you to recall your own first attempt to ride a bicycle. The related thoughts seem to be stimulated because there is some kind of association, or connection between neurons, that leads from the experience to a related memory or thought. Thus, a thought, involving electrical pulses in one set of neurons, may trigger activity in another set of neurons through certain electrical connections. These connections were established previously in *the learning process*, when there was simultaneous activity in the two sets of neurons.

A related thought may be a very different kind of thought than the thought that stimulated it. This suggests that connections are made between widely separated parts of the brain.

Although we don't notice it, it seems to be true that those processes that are conscious at one time are always conscious if attention is directed toward them. Also, we are *never* conscious of most subconscious processes. This suggests that certain portions of the brain are always devoted to conscious activities and other portions are always involved with subconscious activities.

The conscious and subconscious portions of the brain may function in very different ways. One artificial intelligence expert said, "People do seem to have at least two modes of operation, one rapid, efficient, and subconscious, the other slow, serial, and conscious."[2] We will find that these two modes are consistent with the toy model described below.

A TOY MODEL OF CONSCIOUSNESS

Conscious activities may be implemented in the following way. All "conscious areas" of the brain are connected both *to* and *from* working memory. *Output connections* go from conscious areas to working memory, and *input connections* go from working memory to conscious areas. The output connections put conscious mental activities into working memory. The working memory then "transmits" or "broadcasts" the information throughout the conscious parts of the brain by means of the working memory's widespread *input connections* to all of the conscious areas.

Upon receiving information from the working memory, those conscious areas that hold information related to the transmitted information are activated, and *they, in turn, send their output information to the working memory. This new information is then sent out, or "broadcast" to all conscious areas.* In this way, the process generates a number of *mental cycles* of searching all knowledge and memories for *related information* and sending this related information back to working memory. Each mental cycle may require only a fraction of a second.

The whole system needs to be set up so that this process eventually dies down, to keep the brain from entering endless cycles of analysis after receiving new information. This process of updating the contents of working memory in a number of mental cycles makes the toy model share some features with the "multiple drafts model of consciousness" of Dennett.[3] In *The Astonishing Hypothesis*, Crick discusses how the cerebral cortex and the thalamus in the brain may work together to produce a working memory.[4]

The use of the word "information" in the above paragraphs may be misleading. The information may be highly encoded representations of images of objects or sequences of sounds. We don't know how such things are represented within the brain or passed into and out of working memory. Remember, however, that we can design and produce compact disks or other things that can store representations of images or sounds. Remember, also, that the human brain is the most complicated thing in the known universe. We should not be surprised that it has wonderful abilities, or that we are highly ignorant of how it works.

PLAYING WITH THE TOY MODEL

Consider how the system described by the toy model works. *Conscious mental activity puts information into the working memory*. This information is made available, by neural pathways, to other processes such as converting a thought into speech. *Introspection* may be the use of a following mental cycle to examine the contents of working memory from a preceding thought. This may be followed by conversion of the examined memory into long-term memory or speech.

Each piece of new information is inspected and compared with all related information from previous experiences stored in the brain's conscious areas. In this kind of system, *thoughts naturally lead to other related thoughts in the following mental cycles. The implications of a thought tend to be "realized" in the following thoughts*. The related thoughts help set the *context* and contribute *meaning* to the preceding thought.

From the standpoint of a person who is "doing the thinking," *the operation of the system described above seems to behave like our awareness*. Sensory information, *analyzed in detail by subconscious processes*, enters this system and is transmitted throughout the conscious areas of the brain. Following an experience, *the related thoughts give meaning to the experience*. The meaning and all implications of the new information rapidly become present within the brain's conscious areas. Biochemical systems that generate emotional responses are stimulated, so that *the experience has emotional content*.

Together, all these things produce our perceptions of the world, the meanings of events and messages, and the feelings that occur within us. Isn't this a summary of what it means to be aware or conscious? *Although this is not a detailed picture, it seems to describe many of the basic properties of consciousness!*

A remarkable property of consciousness is that it seems so "real." Consciousness is a representation of reality that is constructed by the brain. Why is this representation so realistic that we may mistakenly believe that the reconstructed information is perfect knowledge about the world? One reason is that our nervous system provides us with *a rich flow of highly processed sensory information. It connects this information with a huge amount of related information that gives rich meaning to the incoming data*. Besides the memories of previous experiences, this related information includes *everything that was learned* from our previous experiences.

In the case of vision, we obtain a detailed picture of the surrounding world. The comparison of this image with the brain's stored information makes the picture meaningful. For example, if we see a kitten, our pattern-recognition systems automatically identify a fluffy object of a certain shape as a kitten. The pattern-recognition system has previously been "trained" to recognize kittens. Also, in seeing a kitten, we are immediately aware of the softness of the kitten's fur. We have learned from previous experiences to expect softness when we see the fluffy appearance of a kitten.

Another reason why consciousness seems so real is that the brain constructs a representation that gives a *coherent*, or self-consistent, picture of reality. For example, if you see a bee sting your hand, you soon get confirming information that your hand is in pain. In addition, the brain's representation is a fairly *accurate* representation of reality. A vertical pole *looks* vertical, and you notice if it is tilted. A ninety-degree angle *looks* "square." The brain is well organized, so that we learn from our experiences and construct a realistic representation of reality. *As a product of evolution, the brain must be this way in order to be competent in guiding our actions.*

For us, and according to the toy model, awareness does not need to be stimulated directly by sensory data. That is, a mental process may be a "thought" that arises from the mental activity of previous mental cycles. For example, someone may observe a dog doing something. The observer may realize that this was an intelligent action. This may stimulate the thought that dogs may think. This, in turn, may raise the question of whether animals have souls, leading to the issue of why only people should have souls. The "mental cycling" between working memory and all other conscious brain areas suggests how simple sensations can lead to rather abstract thoughts and the formation of highly abstract concepts.

This iterative process by which conscious parts of the brain reprocess results of previous mental activity results in a brain that is *aware of its own activities*. Thus, the toy model describes what specialists call *a self-monitoring system*. In ordinary speech, of course, the process of reprocessing previous conscious mental activity is simply part of *thinking*.

A *conscious system* of the type described in the toy model might have difficulty dealing with several thoughts at the same time, since the broadcasting of information throughout the system might tie up the whole system. We *do* tend to be limited to one thought at a time. The process of *attention* is necessary for such a system. Attention limits the system to dealing with no more information than it can handle.

Your mind feels as if it is a single entity, even though it results from the functioning of enormous numbers of parts. The intensive communication between all parts of the conscious system and the system's attention to one thought at a time give a strong feeling of the mind's *unity*. The conscious system is organized so that all parts work together on a single process, using a shared working memory. The rigid dedication of all parts to one process gives a secure sense of unity to the resulting consciousness.

Since the purpose of the brain is to control the actions of an organism, it seems that a mind *must* possess this unity. Your brain must function as a single mechanism in order to coordinate your actions. You may be affected by opposing tendencies before taking a course of action, but a single action often needs to follow your decision process, not some compromise between two opposing actions. Since you need a single decision-making system, a unified mind has survival value. Thus, unity is a normal property of a mind that results from an evolutionary process.

The unity of the mind spoken of here refers to activity in the *conscious* parts of the brain. While a conscious process is occurring, there may also be numerous subconscious processes operating in other parts of the brain. For example, the regulation of the heartbeat continues, even while the conscious mind is strenuously engaged with another problem. Sometimes one divides one's consciousness between several tasks, such as driving a car and talking on the telephone. This may be a kind of timesharing, where the brain deals briefly with one and then with the other task. There are limitations of such timesharing. For example, a driver may be slow to react to a crisis if he or she is talking on the phone.

What are you referring to when you refer to "I" or "me" or your "self"? The meaning of these words varies. Sometimes they refer to your entire body, as when you say, "I am getting all wet." Other times the words may be extended to include a car, as when a driver reports, "He hit me on my left fender." Frequently, these words refer to that part of the speaker that feels or thinks, as in "I need a break," or "I want to thank you for a nice dinner." Just what is the real "person" in these last two examples? If it isn't a soul, what is it?

In the "toy model" of consciousness, "I" refers to the functioning of the conscious areas of the person's brain. The *functioning* of these parts of the brain is the mental activity that comprehends the world and itself. In one sense this is the real *person*.

This is rather abstract, but not as nebulous as the idea of a soul. There *is* a physical aspect of a person, since all of a person's memories and mental abilities are embodied within the structure of the person's unique brain. Since much of your personality is due to the properties of your entire body, and your body is the part of the universe that is directly controlled by you, there is a real sense in which your body is part of your self as well.

Since the *person* or *self* arises from the activity of a brain, the self is always located where the brain is located. Your enormous store of memories gives a feeling of continuity to your self, even when many things are changing. Since your self is associated with your brain, there is no reason to worry about your self leaving your body and entering another person's body or wandering outside your body during dreams or when near death. In this picture, near-death experiences in which the self seems to leave the body result from activities of a person's imagination and are related to dreams and hallucinations.

The properties, activities, and characteristics that we have discussed for the conscious system in the toy model are *subjective*. They include *awareness, meaning, emotional content, perceptions, feelings, "seeming real," having abstract thoughts, having one thought at a time, unity of one's mind, and the meanings of "self" and "person."* All of these are experienced separately by each person since they exist separately in each central nervous system. They apply to a single information-processing and storage network: one person's functioning brain. Due to the lack of neural connections between the conscious systems of separate persons, these things are usually completely "private" to the individual, although people may choose to communicate with others and "share their experiences."

There is another way in which mental experiences are subjective. Even if two people are exposed to the same situation, such as viewing the same photographic slide, their experiences will differ. This may be because the persons are in different emotional states when they observe the slide. Another reason is that *the meaning of what is observed will be different for the two people*. This is because the slide will induce different sequences of related thoughts in the two people. This happens because the two had different experiences in the past. The different experiences produced different sets of mental associations (neural connections) in the two people, who then had different thoughts and feelings when they saw the slide.

Even though the experience of viewing a slide may be partly subjective, the two observers would be able to agree on a description of the slide, using words. Words are arbitrary names for a marvelously useful set of *standard concepts*. Words can convert our subjective experiences into messages carrying *objective information* to someone who hears or reads them. Each person is an island of consciousness, but words are bridges between the islands. The lack of language in animals makes it very difficult to study their consciousness, since we usually can't know much about their subjective experiences.

CONSCIOUS AND SUBCONSCIOUS MENTAL ACTIVITIES

What about the *subconscious* brain areas? In the toy model, these differ from the conscious areas by *not being linked to them through connections to and from the working memory*. Since the information in the subconscious areas is not put into working memory, we do not have access to it. We can't think about it, remember it, talk about it, attach meaning to it, or be aware of it. The subconscious areas *can* process information and send it to other subconscious or conscious areas of the brain. If sent to conscious areas, the information may be passed on to working memory, so that it may enter consciousness.

Unlike the conscious system, in which the entire system is tied together to work on a single problem, large numbers of independent subconscious processes can occur at the same time. Each one occupies only a small fraction of the brain's neurons. These processes give the brain abundant computing power by using *parallel processing*. That is, they allow many independent processes to occur simultaneously.

According to the toy model, the processing of information within the brain occurs by *two very different schemes*. The *subcon-*

scious regions do huge amounts of repetitious and rather mechanical processing of information by many independent processes that operate simultaneously. The *conscious regions* combine all their resources to work together on *highly complicated problems* in which information from many sources is brought together, as in interpreting the data obtained from different senses. The brain has achieved greatly improved abilities by applying the more suitable mode of operation to each type of problem.

The toy model suggests that consciousness arose naturally from the particular way in which part of the brain may be organized. This organization, allowing large parts of the brain's resources to be devoted to a single problem, clearly has survival value and is valuable to the organism. *In this model, consciousness occurs in a brain that is organized so that it has a representation of the surrounding world that combines and unifies sensory information with knowledge and meanings obtained from previous learning and experiences.* It seems to be misleading to describe consciousness as an accidental *by-product* of the functioning of the brain. Instead, *consciousness seems to be a subjective effect of having a highly developed functioning brain.* It is *a natural feature* of organisms with reasonably capable brains.

The richness of our awareness suggests that we are far from being the first species to attain consciousness. The earliest conscious organisms may have had quite limited awareness. It is sometimes maintained that consciousness is absent in other animals and that it arose in humans as they gained the ability to speak. This seems unlikely because we are conscious of many things, such as visible scenes, in which the awareness does not seem connected in any way with language or speaking.

When we remember conversations, we can typically recall the meanings of the speaker's sentences, although we are usually not able to recall a speaker's precise words. This suggests that the information is not stored as the actual words of the speaker. Perhaps thinking often takes place without using language. There are many situations in which thinking seems to occur without language, such as when we think about how to arrange pots and pans in a cupboard.

In the toy model, consciousness is a *requirement* for the ability to express oneself by speaking. In this model, the process of composing speech involves translating and vocalizing thoughts that are present in the working memory. We are aware of any thoughts in the working memory and the awareness precedes the translation of the thoughts into speech. In this picture, *meaningful speech would not have evolved before consciousness.*

The toy model suggests that animals that choose to communicate with us are conscious. Other animals may also be conscious, even though they are not able to speak. Interested readers may want to read *Animal Minds*, an interesting and readable book dealing with the possibility of animal consciousness. Its author, Donald R. Griffin, is an American zoologist who is a codiscoverer of *echolocation* by bats. Griffin believes that many animals are capable of thinking and that scientists should actively pursue means of learning about the extent of mental processes in animals.[5]

Dreaming is a sort of spontaneous mental activity that occurs when sensory data channels are "turned off." Imagining and dreaming are both kinds of spontaneous mental activity in which images are produced. It seems likely that animals that can dream can also imagine events that are actually not occurring. Observations of motions of sleeping dogs suggest that they are dreaming. Observations of the so-called *rapid eye movements* are correlated with dreaming in sleeping humans. These movements have been observed in sleeping birds and mammals, suggesting that they dream and, perhaps, imagine things while awake.[6]

The toy model may be mistaken on details. It may even be completely wrong, but it is a way to imagine a real physical system that could exhibit the most basic properties that we associate with consciousness. The toy model shows that we can find objective explanations of at least some properties of conscious systems. The model also explains the presence of extensive areas of the brain that *do not* directly take part in our awareness. In the next two sections we will discuss whether some important aspects of consciousness will always remain mysterious.

DIFFICULTIES IN EXPLAINING CONSCIOUSNESS

Consciousness is a difficult topic to contemplate. As the foundation of one's own reality as a person, it is likely to evoke awe, making one unable to approach it calmly. Because it is so fundamental *to us*, we are apt to overestimate how marvelous and mysterious it is. From the cosmic perspective, however, consciousness is a rare quirk found in a part of that extremely tiny fraction of the universe's matter that happens to be alive. Our biased perspective as conscious beings leads us to bestow a hallowed status on consciousness, though it may not deserve that status.

There are pitfalls waiting for anyone who deals with consciousness. The pathways leading to these pitfalls are downhill and slippery, and we carry a heavy load of cultural beliefs and attitudes that push us toward the pits. One trap is the belief that, *in principle*, consciousness cannot be understood. This trap would have us accept supernatural "explanations" of consciousness. These explanations really explain nothing and lead to a belief in a "ghost in the machine."

A second trap, like the supernatural explanation, is an empty explanation that avoids the main problem. Perhaps the crudest form of this illusory explanation is the idea of a little man in a theater. (The idea is old, from the time when nobody worried about sexism in language.) In this explanation, the brain conveys images and sounds to a little theater within the brain. There a tiny man, or *homunculus*, takes in the presentation. This pushes the difficulty in understanding consciousness from the brain of the person into the tiny brain of the *homunculus*. Is there another even smaller *homunculus* in the brain of the first *homunculus*? This model avoids the fundamental problem of consciousness, but adds nothing to our understanding.

The crudest form of the *homunculus* idea seems too simple, but the idea that there is an especially important part of the brain where consciousness "really happens" is very appealing. This idea may be so attractive because the parts of the brain that are well understood carry out what seem to be mechanical or trivial functions such as detecting straight lines within scenes viewed by a person. This may suggest that the real "seat of consciousness," with its hallowed properties, remains to be found. However, consciousness probably does not arise in a particular *place* within the brain.

In a *mechanical* system such as a clock, certain functions are performed by parts that are located in particular places. The brain, however, is more similar to an electrical system, since it operates by signals transmitted between neurons by electrochemical pulses. In an *electrical* system particular functions may be carried out by complicated circuits that are spread over wide areas. What counts is not the *location* of each part of the system but *how the part is connected to other parts of the system.*

Consciousness requires the linking of results from widely separated parts of the brain since it relates information from all possible sources. It is not surprising that a widely dispersed network carrying such diverse types of information would be extremely difficult for scientists to disentangle and to understand. At present, we can only present plausible and partial solutions to the questions raised by the existence of consciousness.

A real danger exists that correct and adequate explanations of consciousness will not be accepted because of difficulties in accepting explanations that involve *emergent properties*. Emergent properties of matter were mentioned in chapter 5, where they were described as properties that emerge from matter when *special circumstances apply to it, such as the organization of the matter into large numbers of similar units that can interact with each other.* Consciousness may be *the most outstanding example* of such an emergent property. It gives matter a radically new property that is acquired only under very special circumstances. Think of what a tiny fraction of the solar system's matter is conscious!

Sometimes it is hard to imagine *how important* small differences in structure can be in determining the characteristics of emergent properties. Consider an example from physics and chemistry. An atom of carbon has six protons. Nitrogen atoms have seven. The differences between carbon and nitrogen come from nitrogen's extra proton. It causes the atoms of nitrogen to tend to have one more electron than those of carbon, and this leads to the enormous differences in the properties of the two elements.

A difficulty in accepting explanations involving emergent properties is that it is easy to overlook the *special circumstances* and to conclude that mere matter can't be the basis for the remarkable emergent property. In the case of consciousness, one might conclude that, "Mere matter can't produce consciousness." An analogous response would be to look at a stack of lumber, sacks of concrete, wire, pipes, roofing material, and so on, and declare that, "A house is more than just a collection of materials such as these." Of course it is. Besides the materials, a house also involves the careful assembly of the materials according to a detailed plan that specifies the desired final structure.

A similar situation relates ordinary matter to living matter or "life." It is difficult to believe that atoms—mere combinations of electrons, protons, and neutrons—could be brought together to form living beings. Life seems to be so special that it could not be built up from ordinary atoms without adding a "life soul" to the matter. Earlier, however, we saw that living matter is made of ordinary material that is organized in extremely special ways. A nonphysical entity does not need to be added to the material to give it the wonderful properties of living matter. What is needed is a *wonderful organization* of the material.

In the case of a house, *the final product is much greater than the sum of the physical parts*. The same may be said of life. A living person is much more than the raw materials that form his or her body. In the examples of a house and a living person, the *organization* of the materials is, in many ways, the most important part. As a result of the organization of the materials, the emergent property of *livability* comes forth in the completed house. In a person, the organization of the matter produces the emergent property of *life*. In both cases the organization imparts a radically different nature to the materials. This is a close analogy to what I believe is the case with consciousness.

Another difficulty comes from *underestimating how special the circumstances are* that produce a particular emergent property. For example, consider the most intelligent result of billions of years of development of all kinds of life on Earth. We may note that this intelligent organism has a brain that may be the most complicated device in the known universe, involving billions of neurons with trillions of connections to other neurons. These connections were organized by learning processes during tens or hundreds of thousands of hours of experiences. In spite of this, some still say, "Consciousness is too wonderful to be merely the result of the operation of the brain."

You may have a problem accepting the idea that consciousness results from the brain's functioning. It may be hard to believe that all of your memories, experiences, and imaginings result from pulses occurring in a network of neurons. Most people *can* imagine that there are *enough connections* between the neurons to encode all the *information* that forms our memories, abilities, and knowledge. This is probably not the source of the problem.

Much of the difficulty in accepting the working brain as the basis of consciousness comes from the fact that all of this information is encoded in an obscure way. Our ignorance of the code makes it difficult to imagine that the working brain can produce consciousness. In a similar situation, the processes of heredity were quite mysterious before it was understood that DNA carries a code that controls the construction of proteins. The DNA specifies an organism's inherited traits in a way that is *very* obscurely related to the organism's characteristics. Nevertheless, the connection between DNA and our inherited traits is a fact.

In considering the problem of encoding, it helps to note that many information-storage devices, such as videotapes and computer "memories," are encoded. Books are also encoded. Books are written in a particular language that uses arbitrary sounds as words and arbitrary markings on paper to represent words. We are so familiar with the code that a book written in English does not appear to be encoded. Most of us, however, would agree that a book written in Chinese appears encoded.

There is no reason to doubt that connections of neurons *could* encode complicated information. Also, if the encoding of information evolved in order to help organisms survive, then the organisms must be able to *understand* the encoded information and use it to guide their choices of actions. Otherwise, the organism's brains would not be satisfying the need that caused them to evolve into *what they are*: information-processing and storing organs that help the animal survive. So, the fact that the information is coded by means of neural connections is not a problem for the organism itself. Consequently, the fact that the information is encoded should not prevent us from believing that consciousness can arise in a purely material system.

It may seem that we have so many abilities that exceed those of animals that our consciousness must be very different from that of other animals. Much of what gives us such great powers comes from the ability to speak. Earlier, we saw that some apes appear close to the intellectual level at which they could learn to use speech. Considering humanity's evolution from more primitive animals, it seems likely that there was a process of gradual improvement of intelligence and the ability to speak. This process produced human brains that are much more capable than those of our remote ancestors, but the improvements may be only in parts of the brain.

Once, while walking through the Lakeside Mountains west of Utah's Great Salt Lake, it suddenly occurred to me that the experience was similar to that of our distant ancestors who had not developed speech. In such sloping and rocky country, one walks carefully to avoid falling or stepping on a rattlesnake. The experience involves scanning the ground, identifying objects, and maintaining one's balance in climbing over rocks. Information flows into the brain from the eyes, ears, and skin. In this simple activity, intelligence does not seem to be so important, but awareness and alertness are crucial. The experience occurs without speech or thoughts involving words. Consciousness in other animals may be basically like this but with many differences in the strengths of various senses and emotions.

"IT HAS THE RIGHT PROPERTIES, BUT IS IT REALLY CONSCIOUS?"

The toy model suggests how our *awareness* may arise as a result of brain processes. It answers only some of the possible questions, but it helps us imagine that our consciousness is a purely natural process. Suppose that, in the future, we will have a detailed theory of consciousness that explains *everything* about the mind, such as how our memories are stored in the brain, how we have emotions and a sense of beauty, how it is that colors appear the way they do, and how we have religious experiences. Suppose that all of these things are explained in terms of ordinary physical processes in the brain.

Even with such a wonderful theory, many would feel that consciousness is still mysterious. They might say, "The theory explains how the brain produces all the properties of consciousness, but how do we know that it is *like anything at all to be such a system* with all of these properties?" That is, how do we know that a system that exhibits these properties really has any subjective experiences? That the question is asked suggests that it is hard to overcome the effects of our culture's ancient and continuing belief in a supernatural soul. Our history has built the idea of the *mysteriousness* of consciousness into our way of thinking about the world.

There may be no answer to the above question that will satisfy everyone. However, I will try to give a partial answer. I hope that you will try to *set aside your intuitions* about the answer to this question, since these feelings are bound to be affected by the nearly universal belief that our awareness is fundamentally mysterious.

Think of the brain as a system that has the properties listed below:

- Enormous amounts of information are present in the system. The information comes from the senses. Much previous information that entered the system is stored as *memories* and has "educated" the brain by altering connections of neurons as part of the learning process.

- The system uses the process of attention to select a portion of the incoming sensory information for special processing and storage. The selection of important information is done according to *values* or *goals*, some of which were determined genetically and are present in infants, such as the desire to avoid pain.
- When selected information enters the system, the system starts a broad search of memories and knowledge to find related information that gives *meaning* to the new information.
- The entire system participates in a process of *understanding* what the new information shows is happening in the outside world. This is an active process in which the outside world is reconstructed by using the system's sensory and stored information. The system then attempts to predict the future. The predictions are used to make decisions about how to interact with the outside world.

Would it *be like anything at all to be such a system?* The system is an entity that employs networks of neurons to carry out many processes, including those mentioned above and those that are implied by the above list. The system works as a unit to carry out these processes. The system has data (both knowledge and sensory information) about the outside world that it uses to give meaning to, and represent the current situation in, the outside world. All of this knowledge and information is *available throughout this system*, and *this data is given meaning and is understood by the system itself.*

If a system can do what is mentioned above, it must be *aware* of what is happening. In a profound sense, the information and its meaning are *available* within the system in a way that matches what we mean when we say that *it is like something to be that system.* The awareness that is described here is *entirely within the person* since the mental processes occur entirely within the conscious system. Thus, this picture describes one's *subjective* awareness.

To some, this picture of awareness does not explain the *magic* of consciousness. *A useful theory, however, can sometimes explain something that seems magical without requiring that anything really is magical.* In the present case, some magic may be lost in exchange for a big improvement in understanding our own nature.

Even without a wonderful theory that explains everything about the mind, the present knowledge of the brain and such ideas as the "toy model" provide a good basis for believing that consciousness results from the brain's operation.

More detailed models of the mind have been constructed that explain more than has been explained here. The *relational model of the mind* is an example of such a model. It was described by John G. Taylor, a mathematician at Kings College in London, at a recent international scientific meeting devoted to consciousness.[7] This model agrees with some major features of the toy model, such as the process in which meaning is given to our experiences. Taylor believes that many subjective aspects of consciousness can be understood by science, and his model explains many features of consciousness.

CONSCIOUSNESS AS A NATURAL PROCESS

I hope that this chapter and the preceding chapter have helped to make a natural theory of consciousness seem imaginable. Let's summarize the discussion and conclusions.

Modern knowledge of the brain and consciousness supports the idea that *consciousness results from the operation of the central nervous system, especially the brain.* Nothing else seems to be needed to generate consciousness. The brain is an extremely complicated organ, and this enormous complexity seems to be essential for consciousness as we experience it.

Our awareness of the world is not a matter of transmitting sensory signals to a special room within our head where a little person observes what is happening. This idea does not explain consciousness, but only shifts the question from, "How do we have consciousness?" to, "How does the little person in the brain have consciousness?" This is not helpful.

The brain carries out complicated processing of the information that comes from our senses. This processed information is used to *construct* a representation of the world that "makes sense" in terms of the sensory data and what has been learned or remembered from previous experiences.

The representation of the world that is constructed in a brain is not so much a realistic model of the world as it is a presentation of useful information that helps an organism survive. This is so because the brain evolved to choose actions that give the organism and its genes a better chance of surviving.

We are not aware of most information processing that occurs within our brains. Some results of the processing enter the "conscious" part of the brain and become part of our experiences.

I presented a "toy model" that explains some features of consciousness. This model assumes that a *working memory* or *workspace* is linked to all the conscious areas of the brain. The model describes how our thinking leads from one thought to another related thought. It explains an *open-ended process* in which more and more abstract thoughts may be experienced in successive mental cycles.

The model describes how our experiences receive *meaning* by being followed in consciousness by all related material that is learned or remembered from our previous experiences. Information in the working memory is available to us for remembering, for speaking, or for use in constructing the next thought. The model explains why only some brain processes are conscious. *All animals having a brain that includes a portion organized as described here are expected to be conscious.*

The toy model explains the apparent *unity of the mind*, by which one's mind seems to be a single entity, even though the mind results from the functioning of an enormous assembly of neurons. In this model, the most basic *self* or *person* results from the operation of the conscious areas of the brain. The self has a material basis since most of its properties are determined by specific connections between billions of neurons in the brain.

In the toy model, the conscious areas may be spread over much of the brain. What counts in making a part of the brain conscious is not *where* it is located within the brain, but *whether or not it is connected to the working memory.*

If we consider the enormous evolutionary history of the brain, its complexity, and all the experiences that a person has, we should not be surprised by the wonderful capabilities of a person's consciousness.

Eventually, by studying the brain, it will be possible to explain why things seem the way they do to a person, and our consciousness will be well understood. At present, our knowledge is incomplete, but there doesn't seem to be any basic barrier that will prevent us from eventually having a scientific understanding of consciousness. It appears that an adequate understanding of consciousness will involve only natural processes. It will not require a "ghost in the machine" or a supernatural soul.

Many reasons why ancient people believed in the soul are no longer believable. We no longer believe that the soul is something that leaves a person during a dream. The soul is not needed to give motion to the body or to carry out basic bodily processes, such as breathing or metabolism. Earlier, we argued that a supernatural soul is not needed to impart life to a person or animal. In later chapters I will argue that *the human will* operates by natural processes. Consequently, the existence of the will is not a reason to believe in a soul.

I have argued that a supernatural soul is not what enables us to think or be conscious. It also appears that consciousness depends on a very complex physical system, so that consciousness is impossible without a brain or some other physical system that performs the functions of a brain. Since souls, which lack the properties of a complex physical system, are not what are needed to give us continued consciousness after death, they do not seem capable of fulfilling their traditional role of making us immortal. Since souls do not appear to have any function during or after our lives, we can conclude that there is no convincing reason to believe that souls exist. Apparently, the idea of the soul is becoming obsolete. *Is it about time to exorcise the ghost in the machine?*

Although the idea of the soul may be *obsolete* according to many scientists and philosophers, the idea may also be nearly *immortal* because it is supported by major religions. Religions base their beliefs on sacred scriptures, traditions, and decisions by their leaders and members. Scriptures usually don't change, traditions change slowly, and religious leaders and members are often slow to make changes since a stable system of beliefs is a very attractive feature of a religion. Thus, religions are often very resistant to scientific or philosophical arguments. Some religions may see the light and change their beliefs about the soul. Others won't.

Most likely, science will eventually succeed in describing how our consciousness arises from natural processes. It will probably explain how thinking, reasoning, emotions, motivations, and intuition occur as results of activity in the brain, or as results of the brain interacting with the rest of the body and the outside world. Most likely, we, or later generations, will become certain that our mental abilities are not beyond the capacity of material systems. We, or they, will probably learn that it was a great mistake, or *failure of imagination,* to conclude that matter is not capable of supporting such complicated activities. *Eventually, it will probably seem hard to understand how earlier generations could conceive of such complicated activities arising* outside *a system that has billions of functioning material parts.*

NOTES

1. Daniel C. Dennett, *Consciousness Explained* (Boston: Little, Brown & Co., 1991), pp. 264, 270.

2. D. A. Norman, quoted by I. N. Marshall, "Three Kinds of Thinking," in *Toward a Science of Consciousness: The First Tucson Discussions and Debates,* ed. Stuart R. Hameroff, Alfred W. Kaszniak, and Alwyn C. Scott (Cambridge: MIT Press, 1996), p. 730.

3. Dennett, *Consciousness Explained,* pp. 111–38.

4. Francis Crick, *The Astonishing Hypothesis: The Scientific Search for the Soul* (New York: Simon and Schuster, 1995), pp. 240–42.

5. Donald R. Griffin, *Animal Minds* (Chicago: University of Chicago Press, 1992), p. 260.

6. Ibid., p. 258.

7. John G. Taylor, "Modeling What It Is Like to Be," in *Toward a Science of Consciousness,* pp. 353–76.

11

CONSCIOUSNESS AND GHOST-BUSTING

ECCLES'S MODEL OF THE SELF AND THE BRAIN

MORE THAN A CENTURY AGO, Charles Darwin maintained that thinking is an entirely natural process that will eventually be explained by his theory of evolution.[1] At present, the idea that consciousness results entirely from the brain and does not require a soul is widely accepted by scientists working to understand the mind. However, *one authoritative scientist did not share this opinion.* An unusual model of the brain's role in consciousness has been described by Sir John C. Eccles (1903–1997). He was an Australian and a Nobel Prize winner in physiology. A few years ago, he presented his views in the book *How the Self Controls Its Brain.*[2]

Eccles presented a dualist theory of the mind. He believed that *our mental processes exist in a separate, nonmaterial world.* The "self" is nonphysical in his model, and is equivalent to what, in religious terminology, is called the *soul.* He believed that perceptions, thoughts, feelings, and the self all exist within a nonphysical "World 2" (his name for the spiritual world). He believed that the brain communicates sensory data, through millions of communication channels in the brain, to the self. The self has experiences and thoughts and makes decisions. The results of decisions are communicated from the self back to the brain, where the nervous signals that produce bodily actions are generated.

As a proposed scientific explanation of human nature, Eccles's model is an unusual hybrid in that it mixes supernatural properties of the soul with natural properties of the body. In this model, consciousness occurs within the soul and is not explained, but is assumed to be a power of the soul. There is no explanation of where the soul gets these powers. The soul is presumably created by an act of God, while the properties of the body are inherited by means of DNA. Besides requiring the existence of a soul, the model requires mechanisms for two-way communications between the soul and the brain.

The main innovation of Eccles's model is the means by which he believes the soul communicates with the brain. Within the axons of neurons, signals are transmitted by electrochemical pulses. One of Eccles's outstanding accomplishments in physiology dealt with the transmission of neural signals at the synapses *between neurons*. At connections between neurons, signals are carried by *transmitter chemicals* or *neurotransmitters* that are dumped into a narrow gap between the signaling neuron and the following neuron. The neurotransmitters are initially contained in small packets on the signaling neuron's side of the gap. When a pulse occurs in a signaling neuron, a packet of the neurotransmitter is sometimes discharged into the gap.

The individual packets are so small that the emission of chemicals from the packets may be slightly affected by some random motion due to quantum mechanical processes, discussed in chapter 13. Eccles proposes that the soul, through a quantum mechanical process, can bias the packets toward emptying or not emptying their chemicals into the gap between the neurons. Since there may be as many as one hundred thousand synapses involved in the most elementary brain processes, Eccles maintains that tiny quantum mechanical effects in all of these synapses could alter the outcome of a process. By this means the soul could control the brain and thereby control the movements of the body.

Eccles presented experimental evidence showing that *attention to certain tasks is associated with enhanced physiological activity in certain parts of the brain*. He suggests that this evidence shows that the *mental (that is, nonphysical) process of attention produces physical effects in the brain*. A theory of attention by David LaBerge was discussed in chapter 8. That theory described brain circuitry that may implement attention. If LaBerge's model is right, attention would *have to be* accompanied by activity within the brain *since it is a brain process*! So, the evidence presented by Eccles can just as well be interpreted as showing that *attention is a brain process* rather than a function of the soul. Eccles did not mention this alternative possibility.

I have major objections to Eccles's model. One is that the model is heavily oriented toward a view of the mind that is supported by a religious belief, but which is not supported by any scientific evidence. Other scientific theories do not assume that part of what is being explained is supernatural or miraculous. In my opinion *there is no good reason to believe that the mind needs a supernatural explanation*. From this point of view, the introduction of unsupported entities (the soul) and unsupported interactions (two-way communications between the brain and the soul) is unnecessary and unscientific.

Eccles maintains that his model is consistent with physics because the quantum processes that affect the emission of neurotransmitters do not violate the law of conservation of energy. Of course, all of physics is not contained in this one conservation law. Eccles does propose that there is an interaction that is previously unknown to science: the influence on the brain by the soul. *This proposal is in disagreement with (or at least outside) known physics*, and Eccles presents no strong evidence to support it.

I also disagree with the Eccles model's use of *random* quantum processes to implement decisions that are already *determined* by the soul. According to quantum physics, the randomness in quantum processes is truly random. That is, the individual quantum processes are unpredictable in principle, and would not be affected by a person's previous mental decision. The soul's decisions are presumably not random, so any quantum behavior by which the soul controls the brain *would not be truly random* behavior. So, Eccles is asking for more than new kinds of interactions to be introduced into physics: he is also suggesting a major modification of quantum mechanics. These significant revisions of physics would be justified only if they were supported by very important reasons, which I believe are absent.

Finally, I would like to present a partly facetious argument against Eccles's model, based on evolution. The human brain is much larger than the brains of human ancestors of more than two million years ago. Such a large brain has serious disadvantages as far as natural selection is concerned. It uses a great deal of energy, contributing to our need to obtain large quantities of food. The large brain is related to the large size of the heads of infants, making human births a more dangerous process than the births of other animals. By Eccles's model, evolution should have pushed humans toward having *smaller brains that would merely serve as interfaces with the soul*. Then the real work of thinking could have been done "for free" by the soul, reducing the costs of human intelligence in the physical world.

Unlike the outcome described above, the human brain is very large and has the disadvantages that were mentioned. It appears that the brain needed to grow large in order to support the great intellectual capacities of normal humans. The "free lunch" option of having the soul do all of the serious thinking did not develop. I maintain that this option was not possible, since there was no supernatural entity available to carry out this function.

MINDS, COMPUTERS, AND ROBOTS

From the toy model one might think that consciousness is so simple that it could be produced on one's home computer. This is far from the truth. The toy model is simple, but many complications were ignored in describing it. One major complication is that the model dealt with a system that already possesses the ability to represent the world in a way that is meaningful to the system itself. If an artificial system is to be similar in ability to a human being, the system must possess memories and knowledge equivalent to what a human obtains from years of experiences. The system would have to possess many thousands of learned concepts and skills in order to function at the level of a human being.

In order to acquire concepts and skills that are at all similar to those of a human being, the system would need to have senses such as sight and feeling in order to interact with the world and learn from its experiences. Since much essential learning would require manipulating and touching objects that may not be "near at hand," the system would benefit from having the ability to move itself and to handle objects. So, the system would seem to need to be both a "robot" and a computer with an enormous ability to process the large flow of data from its "eyes" and "hands."

The robot could be the "body" of the system. It could be connected by a radio transmitter to a powerful but stationary computer (the "brain"). Thus, the system might resemble the human in the facetious argument of the previous section, in which the body would have little brainpower, but would have a million or so channels connected to the "soul" that would have all the computing power.

The robot would need to keep its battery charged at all times. It would also need to "want" to explore the world in order to learn from the world. In other words, from the very start of its operation, it would need to have "motives" that would lead it, for example, to seek a battery charger when it needed charging. It would also need to be motivated to obtain knowledge about the world when it was well charged. These motives would need to be programmed into the system from the start, just as a newborn baby tends to want food when it is hungry. In practice, these motives would amount to a built-in tendency to take action to satisfy the highest priority "need" that had not been fulfilled.

Perhaps the robot would also need a "sleep" mode in which it could take care of certain brain processes that are not compatible with the brain's activities when it is awake. The robot's brain would need to be set up so that it would be able to use its experiences to learn about three-dimensional space and the properties of objects. It would need to be preprogrammed to want to learn how to move about. In a human, of course, the code for all the preprogramming is contained within the person's DNA. If the robot would be intended to communicate with humans, it would need to have hearing and special software, hardware, and training in order to be able to learn to understand speech and to speak.

It is not hard to summarize what *might* be needed in the robot's brain. The robot might need to be able to carry out all the brain processes of a newborn baby and have all the ability to learn and continue to develop that a human baby has. Perhaps the robot would not require *all* the brain functions of a baby in order to be able to develop a mind that would seem indistinguishable from a human mind. Some shortcuts might be possible since the robot might have a body that is considerably simpler than a human baby's body. On the other hand, it might not be possible to take many shortcuts if the robot were expected to be emotionally and intellectually indistinguishable from a human.

At some time during the training process, the robot would have gained enough experience that it would be able to make sense of, and attach meaning to, what it sees in the world. Perhaps at this point it would become conscious. The development and training process for such an artificial intelligence would be extremely expensive. It might be much more expensive than the first trip to the Moon. All the results of the training, however, would be contained within the artificial brain; that is, a powerful computer. *A second robot could be "trained" just by copying what is stored in the brain of the original robot!* So, an unlimited number of robot clones could be produced at a tiny expense compared with the cost of designing and training the original robot.

It may seem that this section has drifted into the realms of science fiction. How realistic is the possibility of a conscious robot? First, I should admit that we do not know nearly enough about the functioning of the brain that we would be able to program the computer to duplicate the operation of a real brain. So, this whole discussion is about the distant future. *If civilization continues to make scientific progress, there is no reason to think that there are any fundamental limits to what may be learned about the functioning of the brain.* Eventually the necessary knowledge should be available to carry out the artificial-consciousness project described above.

Let's suppose that sufficient knowledge about the structure and functioning of the brain is available. Could a brain be replaced by a computer? Let's approach this question by starting at the level of an individual neuron in the brain. Could the computer duplicate the operation of a neuron? I think that it could. A simulated neuron in a computer program could take numerous inputs from other simulated neurons and produce outputs that would resemble the outputs of real neurons to a reasonable degree of accuracy.

The output of a simulated neuron might be a string of numbers representing voltages at some position on the neuron's axon for each time interval of, say, a thousandth of a second. The computer could take outputs from an original simulated neuron and apply them to the inputs of all the other simulated neurons to which the original neuron is supposed to be connected. This could be done for all the simulated brain's neurons, duplicating the activity of a real brain. The computer could divide its time between different neurons or it might use parallel processing to simulate different neurons simultaneously.

The simulation process would need to include effects of learning and memory processes that alter the connections between neurons. The effects on the functioning of the neurons by various chemicals within the brain would also need to be simulated. This would duplicate effects of emotions or other influences on the neuron's environment. The goal of all this would be to produce the same patterns of activity in the duplicated neurons in the simulated brain as in the neurons in a real human brain. Our present knowledge suggests that the brain activity that produces consciousness consists of sequences of pulses in the neurons of the brain. To reproduce the connections between neurons and the pulse activity in one hundred billion neurons of a real brain is obviously an enormous task!

The main goal of the robot program might be to produce behavior similar to that of a human. That implies that all the numbers being shifted around in the computer would need to have results in the real world. In a human brain, results of brain processes are used to send pulses to motor neurons that control muscles. In the robot, the simulated decisions to take action could be carried out by converting simulated pulses in simulated motor neurons into voltage levels that would control electrical motors. Thus, the robot could move about under the control of its artificial brain.

Now suppose that all of this work were completed. Would the robot exhibit human intelligence and consciousness? In the toy model, many of the properties of consciousness arise from the structure of the conscious system. The robot would possess the same kind of functional structure that is present in a human brain, but implemented in an electronic computer rather than a brain. The robot would also possess the knowledge and skills gained by extensive training that copied the early experiences of a human being. It would seem likely that it would be intelligent and conscious.

A majority of scientists working on issues related to the mind would probably agree that such a robot could display all the features of human intelligence and that it would be conscious. One prominent scientist, however, has presented lengthy arguments that conclude that at least some aspects of human thinking processes could not be completely duplicated by computers. He is Roger Penrose, a British mathematician and cosmologist. His ideas have been presented in his 1989 bestseller, *The Emperor's New Mind*,[3] and his more recent book, *Shadows of the Mind*, published in 1994.[4]

Penrose argues that some results obtained by mathematicians could not be obtained by computers, even in principle. Consequently, he maintains that computers cannot duplicate at least part of what happens in the brain. He also notes that all the known physical processes can be simulated by computers. He concludes that *the functioning of the human brain must involve some processes that go beyond the known laws of physics*. He is not a dualist. He does not believe that supernatural effects are involved in the functioning of the brain. He suggests that the new processes might arise in an improved theory of gravity.

Frank Wilczek, a prominent American theoretical physicist,

reviewed Penrose's *Shadows of the Mind* for the publication *Science*. Five points in Penrose's arguments were not convincing to Wilczek.[5] Penrose himself anticipates that many scientists will not accept his ideas and that they will search for loopholes in his arguments. He does not regard such an approach as unreasonable.[6] It should also be noted that Penrose does not rule out the possibility that intelligent devices could be built. He maintains, however, that the devices would have to go beyond the capabilities of electronic computers and make use of the "non-algorithmic processes" that he believes are present within the human mind.

A RECENT ARGUMENT FOR THE SOUL

A group of scientists and theologians recently wrote and edited a book advocating a novel set of beliefs about the soul. Although they abandon traditional dualism, they still attempt to preserve a special status for the soul, consistent with what they regard as the essential tenets of Christianity. Their book is *Whatever Happened to the Soul: Scientific and Theological Portraits of Human Nature.*[7] The book's chapters, though written by different authors, maintain a consistent perspective. These theologians and scientists have been remarkably flexible in painting a portrait of the soul that, on the surface, seems acceptable to science as well as theology.

My conclusions about the soul differ from those of that book's authors. There are some points of agreement, however, as in the following, written by Nancey Murphy:

> To sum up, science has provided a massive amount of evidence suggesting that we need not postulate the existence of an entity such as a soul or mind in order to explain life and consciousness. Furthermore, philosophers have argued cogently that the belief in a substantial mind or soul is the result of confusion arising from how we talk. We have been misled by the fact that "mind" and "soul" are nouns into thinking that there must be an object to which the terms correspond. . . . When we say a person has a

mind, we might better understand this to mean that the person displays a broad set of actions, capacities, and dispositions. Authors of some of the following chapters make a parallel move with respect to the soul.[8]

Let's look at some basic disagreements between what I will call "this book" (*Are Souls Real?*) and "that book" (*Whatever Happened to the Soul?*). A major disagreement concerns "causal reductionism," which is denied by that book's authors. As they define it, "*Causal reductionism* is the view that the behavior of the parts of a system (ultimately, the parts studied by subatomic physics) is determinative of the behavior of all higher-level entities" (p. 129).

Unlike the authors of that book, I support causal reductionism, as they have defined it. In chapter 5, I discussed the subatomic particles and the actions between them. Near the chapter's beginning, I noted, "Our world operates by means of orderly interactions of the basic building blocks of nature. Science has not uncovered any exception to this rule. The building blocks, called *elementary particles*, are described in this chapter." In these sentences, I mean to imply that the future motion and behavior of all matter in the universe, including living and thinking matter, results from the motions of, and the physical interactions between, the subatomic particles (the parts) that make up this matter.

Unlike those authors, I believe that the behavior of large-scale objects is completely determined by the behavior of the subatomic particles that make up the objects. The behavior of the particles, in turn, is a result of the forces exerted on each particle by all other particles. The subatomic particles make up atoms, which make up larger-scale objects. Evidently, I disagree with those authors and support causal reductionism, as they define it.

Let's consider some examples of how large objects obey causal reductionism. Suppose the atoms that make up a block of wood go one way. Could the block go another? Could the atoms in your tongue go one way and the tongue go another? Of course not. Let's look at another example, closer to religious concerns. Suppose a person needs to choose between doing a good act or an evil act.

From the viewpoint of that book's authors, as well as my own views, the mental activities involved in making the decision occur in the brain. In principle, if not in practice, shouldn't science be able to describe the actions of all the brain's neurons that are involved in making the decision? Don't these neurons behave in an orderly fashion, governed by scientific laws? Except for small random motions due to "quantum fluctuations," discussed in chapter 13, don't these laws of nature specify what the behavior of all the neurons will be? Don't all the neurons, by their collective behavior, determine the outcome of the decision process? If you agree with this line of thought, then you, too, seem inclined to be a causal reductionist, unlike the authors of that book.

As a physicist, I maintain that a complete description of a physical situation, even a person making a decision with moral consequences, or people interacting in a social situation, can be done at the bottom level. That is, the *complete* specification of the situation can be done by describing the physical behavior of all the fundamental particles, or other very microscopic objects, that are involved. By taking into account the forces acting on each particle due to all the other particles, the behavior of each particle can be understood. The "completeness" of this specification implies that one can describe a situation adequately without taking into account "top-down" effects of the type discussed below. Top-down effects, however, are essential to the picture of the soul supported in that book.

Nancey Murphy defines "top-down" causation as taking "account of causal influences of the whole on the part, as well as the part on the whole" (p. 130). From my perspective, these influences are already included when one considers all of the interactions between all of the fundamental particles. If the interactions between particles are all that is meant by top-down effects, then the top-down effects are equivalent to the interactions between the basic particles. In this case, these effects do not really introduce anything very remarkable, and they pose no problems for science and there is no reason to deny causal reductionism. I believe, however, that this is not what the authors of that book want, since they feel they must reject causal reductionism.

If, on the other hand, the top-down effects are not equivalent to interactions between the basic particles, then *they add something to physical processes that is outside, or contrary to, current scientific knowledge*. This may be what the theologians want. I will argue below that, in the case of that book, the top-down effects are a way of introducing supernatural influences into physical situations. This is acceptable to religion, of course, but not to science.

The perspective presented above is strongly influenced by the fact that I am a physicist. It is quite possible even for scientists who are not physical scientists to overlook such arguments. Based on the "Contributors" section of that book, none of that book's authors are physical scientists.

I should note that there are many forms of reductionism, and some extreme forms have a poor reputation. I am not an extreme reductionist. I do not believe that higher-level sciences, such as psychology and sociology, will ever be made obsolete by being reduced to physics and mathematics. I regard higher-level abstractions such as morality and love as meaningful topics of discussion. I do believe, however, that these topics can be meaningful without implying the existence of any supernatural effects.

In *Whatever Happened to the Soul*, Nancey Murphy lists two issues related to "our traditional conceptions of personhood" that lead the authors of that book to reject causal reductionism and to emphasize top-down effects. She describes the first issue as follows:

> First, if mental effects can be reduced to brain events, and the brain events are governed by the laws of neurology (and ultimately by the laws of physics), then in what sense can we say that humans have free will? Are not their intendings and willings simply a product of blind physical forces, and thus are not their willed actions merely the product of blind forces?[9]

Before replying, let me note that I will discuss free will later in chapters 13 and 14. I will consider whether all forms of the free-will idea are compatible with scientific laws. I will reject some forms that depend on "minor miracles" (certain supernatural effects). As most readers will expect, I *do* maintain that "mental events can be reduced to brain events," and that "brain events are governed by the laws of neurology (and ultimately by the laws of physics)."

Since Nancey Murphy advocates a form of free will that does not allow brain events to be governed by the laws of neurology, I feel that her views about free will are probably mistaken. As I will discuss in this book's later chapters, the conflict between the laws of neurology and free will arises from a belief in a particular form of free will. This form of free will, which I will call "Objective Free Will," implies that supernatural effects are everyday occurrences.

At the level of individual neurons, the brain's decision-making process does appear to be "blind." If, however, we consider the entire ensemble of neurons that are involved in making a decision, they are interconnected with each other in such a way that *the ulti-mate decision takes into account all of a person's wishes, values, motivations, and reasons.* These motives are the result of connections that have been made between neurons during the previous development of the person's brain. From my perspective, our *consciousness* of our wishes, values, motivations, and reasons strongly suggests that these things are involved in the actual decision process. Contrary to Nancey Murphy's implication, there is no reason to believe that the decision process, as a whole, is "blind," even though the individual neurons have no understanding of the effect of their activity on the entire conscious process.

That book's discussion of the two issues continues:

> Second, if mental events are simply the products of neurological causes, then what sense can we make of *reasons?* That is, we give reasons for judgments in all areas of our intellectual lives—moral, aesthetic, scientific, mathematical. It seems utter nonsense to say that these judgments are merely the result of the "blind forces of nature."[10]

By means of evolution, the "blind forces of nature" created vision, imagination, and creativity! I hope that readers of this book will be able to imagine, and to some extent understand, how our brains produce all aspects of human consciousness, including our *reasons* for making certain judgments or decisions.

Nancey Murphy's second issue implies that it is inconsistent to believe: (1) that we make judgments by considering reasons, and (2) that the judgment-making process is done by a vast network of neurons. I believe, however, that the two statements are merely different ways of describing the same process.

Among the brains of all species on Earth, the human brain may be unique in that it evolved a remarkable ability to be aware of its own processes. For example, after making a judgment or decision, we can analyze the process and describe the motives and causes that supported or opposed the actual judgment or decision. In other words, we are aware of, and can describe, the apparent *reasons* for and against the actual outcome of the mental process.

From a neurological perspective, we can analyze the same mental process at a lower level. We can describe how enormous groups of neurons, like sign-carriers at a political party's national convention, carry signals that support a particular judgment or decision outcome. In referring to the combined activity of all the neurons that support a particular motive, we may simply call it one of our motives or reasons. These "reasons" that Nancey Murphy speaks of are a shorthand way of naming, with one word, the collaborative activity of thousands of neurons. The next chapter will make it much clearer how motives arise and compete within the enormously complicated neural networks of the brain. At this point, it should be obvious that being able to list reasons for, and against, our judgments is not inconsistent with the fact that judgments are the result of "neurological causes." Although the same mental process may seem very different when described in ordinary speech rather than in terms of the detailed physiology, the two descriptions are not inconsistent. They are different portraits of a single reality.

Although it is not easy to understand how reasons are represented in a network of neurons, think of how difficult it is to understand how reasons exist *outside* the brain! Are our reasons registered in some kind of spiritual scratch pad? Presumably, the details of how reasons could be maintained *outside* the brain are to be regarded as an eternal mystery.

After raising the two issues discussed above, Nancey Murphy writes:

> If free will is an illusion and the highest of human intellectual achievements can (*per impossible*) be counted as the mere out-working of the laws of physics, this is utterly devastating to our ordinary understanding of ourselves, and of course to theological accounts, as well, which depend not only on a concept of responsibility before God, but also on the justification (not merely the causation) of our theories about God and God's will.[11]

In later chapters I will discuss why I believe part of the free-will idea is an illusion. I will discuss the consequences of this belief, and I will leave it to readers to decide whether my views are "utterly devastating to our ordinary understanding of ourselves." Perhaps Nancey Murphy's statement, together with the viewpoints presented in this book (*Are Souls Real?*), should put to rest the idea that there are no longer any major disagreements between science and theology.

Let me summarize my discussion of *Whatever Happened to the Soul?* That book's authors separate themselves from much conventional religious thought by opposing dualistic views of the soul. I agree with them on this. In their attempt to preserve some role for the soul, however, they introduce a scheme that denies causal reductionism and emphasizes certain mysterious top-down effects. Nancey Murphy defines causal reductionism as "the view that the behavior of the parts of a system (ultimately, the parts studied by subatomic physics) is determinative of the behavior of all higher-level entities." She argues that if causal reductionism is true, it destroys theological ideas about free will and how *reasons* determine our judgments.

In disagreement with that book, I maintain that causal reductionism is likely to be true. *Denying it* implies that human nervous systems do not operate according to the laws of physics that are supported by numerous experiments and are thought to apply to all matter. Religion, rather than science, appears to govern the thinking of that book's authors. By my thinking, that book's ideas about the soul are likely to be mistaken. That book illustrates how difficult it is for well-educated and scientifically aware theologians to bring traditional religious ideas about the soul into harmony with their scientific knowledge.

INTRODUCING THE "NATURAL SOUL"

Science will probably succeed in describing how our consciousness arises from natural processes. It will probably explain how thinking, reasoning, emotions, motivations, and intuition function as a result of the activity of the brain, and as a result of the brain interacting with the rest of the body and the outside world. Most likely, we, or future humans, will learn that human mental abilities are not beyond the capacity of material systems. We will probably learn that it was a great mistake, or *failure of imagination*, to conclude that matter is not capable of supporting such complicated activities. *Eventually, as mentioned in the previous chapter, it will probably seem hard to imagine how earlier generations conceived of such complicated activities arising* outside *a system that has billions of functioning material parts.*

For the present, however, anyone who denies the reality of the soul faces the problem that many people within our current culture confuse the existence of the mind and other capacities with the existence of the soul. As a result, many people tend to believe that denial of the existence of the soul amounts to a denial of the existence of many of the higher human capabilities. Perhaps we need a new term to represent those real, higher capacities of humans that are commonly included within the supernatural soul, but are not really supernatural in origin. I suggest that we call this bundle of higher capacities the *natural soul*.

The natural soul includes the mind, the will, and one's conscience. It includes one's abilities to think and make moral decisions. Since the soul is also commonly associated with our emotions, *the natural soul should also include emotions and feelings.* The limitations of our ideas of free will will be discussed in chapters 13 and 14. The natural soul is meant to include only those parts of the complicated idea of human free will that are consistent with science. The natural soul is not expected to be immortal, since that seems contrary to the idea that the abilities connected with the soul result from the normal functioning of one's nervous system and body.

The "natural soul" covers a broader set of abilities than the "mind." It may refer to all human psychological processes, while making it clear that one considers them to be *real and completely natural processes.*

NOTES

1. Gerald M. Edelman, *Bright Air, Brilliant Fire: On the Matter of the Mind* (New York: BasicBooks, 1992), pp. 42–51.

2. John C. Eccles, *How the Self Controls Its Brain* (New York: Springer-Verlag, 1994).

3. Roger Penrose, *The Emperor's New Mind: Concerning Computers, Minds, and the Laws of Physics* (New York: Oxford University Press, 1989).

4. Roger Penrose, *Shadows of the Mind: A Search for the Missing Science of Consciousness* (Oxford: Oxford University Press, 1994).

5. Frank Wilczek, review of *Shadows of the Mind: A Search for the Missing Science of Consciousness*, by Roger Penrose, in *Science* 266 (9 December 1994): 1737, 1738.

6. Penrose, *Shadows of the Mind*, p. 202.

7. Warren S. Brown, Nancey Murphy, and H. Newton Malony, eds., *Whatever Happened to the Soul? Scientific and Theological Portraits of Human Nature* (Minneapolis: Fortress Press, 1998).

8. Nancey Murphy, ibid., p. 18.

9. Ibid., p. 131.

10. Ibid.

11. Ibid.

12

CONSCIOUSNESS AND FEELINGS

I N EVERYDAY CONVERSATIONS, FEELINGS AND motivations are often attributed to one's soul. Even someone who believes the brain generates one's thoughts may have trouble imagining that feelings are produced by a network of neurons. Can feelings as diverse as pain and pleasure, curiosity, fear, anger, guilt, and happiness be explained by ordinary physical processes? This chapter aims to show that this is possible.

The Toy Model describes a physical system with awareness, but it neglects the role of feelings and emotions in consciousness. The model portrays a kind of "cool intelligence," which could announce, in a flat, mechanical tone, "Goodbye. Have a nice life." after learning that it was about to be shut down permanently. Any conscious system that acts in the real world, however, needs something similar to our *values* or *goals* in order to weigh different options and make decisions about what actions to take. Even the mental process of attention requires built-in criteria that specify what is interesting and what is to be ignored. In humans and some other animals, various intrinsic biological goals are enforced by such feelings as pain, pleasure, and the emotions.

Our consciousness is usually affected by our feelings. This

chapter attempts to explain why a conscious animal needs feelings, and how they are produced. I will argue that our feelings and motivations evolved to help us survive and reproduce. Our pains, pleasures, and emotions are the means by which the goals of biological fitness steer our actions. *By complicated processes, these primitive goals control much of our behavior.* The processes are so effective that, most of the time, our amazingly developed mental abilities are at the service of these primitive drives that we share with other animals. Unlike other animals, our brains allow us to serve these goals by very indirect means. As a result, even the connections between our own behavior and our motivations can be hard to trace.

Feelings are poorly understood, and much of this chapter is quite speculative. Some outstanding books have dealt with related issues, giving me courage to proceed. Especially helpful are *Descartes' Error: Emotion, Reason, and the Human Brain*, by Antonio R. Damasio;[1] *Bright Air, Brilliant Fire*, by Gerald M. Edelman;[2] and *The Emotional Brain: The Mysterious Underpinnings of Emotional Life*, by Joseph LeDoux.[3]

"Feelings" are interpreted broadly here, including emotions, motivations, and sensations. This chapter will concentrate on emotions, pain, and pleasure, with a brief discussion in the next section of some questions raised by our perception of color.

WHY DO COLORS LOOK THE WAY THEY DO?

The electromagnetic spectrum extends from radio waves to high-energy gamma rays. At the lower end of the AM dial, radio wavelengths are about one-third of a mile, while some energetic gamma rays from outer space have wavelengths less than a billionth of an atom's diameter. The entire spectrum spans a wavelength interval stretching more than seventy factors of two, but we *see* only a narrow interval of less than *a single factor of two*. Within this narrow window, however, we can distinguish violet, blue, green, yellow, orange, red, and many other hues.

Evolution's choice for the spectral window of human vision is reasonable, although the selection was not done by reasoning. Our eyes evolved to detect solar radiation reflected by objects on Earth. The Sun produces radiation ranging from radio waves to gamma rays. Much of this radiation is absorbed by the atmosphere. Since the absorbed radiation doesn't reach ground level, it isn't useful for vision. Visible light and radio waves, however, *do* penetrate the atmosphere. Radio wavelengths are too long for high-resolution vision, unless the organism has "eyes" the size of microwave dishes! This leaves visible light as the likely wavelength interval for vision. Light is also the most intense form of solar radiation, so it is the obvious choice.

Among mammals, primates excel in distinguishing colors. It is easy to imagine why our ancestors developed color vision. Most primates eat fruit and live in trees. Color vision is extremely useful for distinguishing ripe fruit against a background of leaves and branches.

Of course, trees coevolved with animals with color vision, and it was advantageous for trees to have colored fruit. This made it more likely that birds and primates would find and eat the fruit, spreading the seeds throughout the forest.

Some fruits change color as they ripen, so it is very useful for primates to distinguish *different* colors, so they can learn to pick only ripe fruit. Thus, primates can distinguish red, orange, yellow, blue, violet, and green, not just "tree" and "ripe fruit."

To make it possible for ripe fruit to be distinguished from other objects, distinct wavelengths need to be displayed differently in the representations of the outside world constructed by primate brains. A similar requirement occurs when various materials or objects need to be portrayed in black-and-white drawings. In geological maps, mechanical drawings, and black-and-white cartoons, areas can be distinguished by representing them by various patterns of dots, closely spaced slanted lines, and cross-hatching. In such cases, we know that the patterns are artificial, so there is no temptation to believe that they are intrinsic characteristics of the portrayed objects.

In color vision, however, we have built-in systems for distinguishing and representing differences of hue. By subconscious processes, our brains handle our sensory cues automatically, and we only become aware of different colors within different patches of our visual field. Rather mysteriously, the colors appear, with no suggestion that they result from a complicated detection and reconstruction system. Because the neural operations that differentiate and display different hues are *subconscious*, we tend to believe that colors are really "out there" in the physical world.

Of course, different-colored objects normally *do* have different physical properties. We can sort objects by color because the light coming to our eyes from the different objects has different wavelengths. More precisely, light from the different objects has different *distributions* of wavelengths, so that it produces differences in the relative responses of the three types of light-sensitive *cone cells* in our retinas. We experience wavelengths differing by 15 percent as *entirely different colors. Our visual system converts modest wavelength differences into major qualitative differences in what we perceive.* We do not know the details, but this process is almost certainly a function of specialized brain circuits. In principle, there is nothing mysterious about distinguishing small wavelength differences and representing them by entirely different neurons. By such a process, small *quantitative* differences can be converted into clear *qualitative* distinctions in what is perceived. We call these qualitatively different outputs "colors."

The brain represents different colors in an arbitrary way, just as a black-and-white cartoon uses arbitrary patterns to represent different colors. Some philosophers have read a deep significance into this. From the perspective presented here, the "other world of subjective experience" in which different colors are represented in our consciousness has about as much reality as the "other world" of all possible chess games. Both of them are products of complicated brain processes. They are significant only to some life-forms on Earth. They are meaningless outside of Earth: that is, almost everywhere!

Different colors do, of course, have different psychological effects. Perhaps because of our evolutionary origins, the colors that identify ripe fruit are also the bright and pleasant colors commonly used in children's candies and toys. We may have innate tendencies to prefer these colors. As we mature, we attach emotional values to colors as a result of our experiences. For me, red has complicated emotional overtones. Besides ripe fruit, it is connected with glowing embers, raw meat, and blood. Orange, however, seems relatively innocent, and I associate it with the fruit. For someone in Northern Ireland, however, orange may incite strong feelings.

Our neural structures and connections probably account for all our perceptions and feelings produced by colored objects. A reader may object that I have not discussed why we *feel* anything at all. Answering this is a major objective of the rest of this chapter.

FEELINGS AS PHYSICAL PROCESSES

Feelings are the effects, in one's consciousness, of complicated physical processes occurring throughout the body, especially in the brain. That we are aware of our feelings, and that certain thoughts trigger feelings, shows that the brain's conscious areas are involved. That our feelings "come to us," and sometimes even surprise us, suggests that they originate partly within subconscious areas of the brain or outside the brain. Since we often feel the effects of emotions in parts of the body outside the brain, the body must play a role in our feelings.

Although emotions have often been viewed as subjective states of consciousness, other people can detect a person's emotions by his or her body language, including facial expressions, and such physical responses as blushing, sweating, and trembling. Even a machine, the polygraph or "lie detector," can detect subtle emotional responses by measuring and displaying the autonomic nervous system's effects on respiration, blood pressure, pulse rate, and sweat-gland activity.

Feelings can be triggered by electrical activity in the brain. Evidence for this comes from electrical stimulation of regions in the brain's *temporal lobes*, done to find problem areas in patients suffering from epilepsy. The procedure was perfomed during brain surgery on fully conscious patients, who were not suffering from pain or seizures during the procedure. The results were as follows:

Excitation of certain parts of the temporal lobes produces in the patient an intense fear; in other parts it causes a strong feeling of isolation, of loneliness; in other parts a feeling of disgust; and in others sorrow or strong depression. Stimulation of some parts causes a feeling of dread rather than of fear, a dread without object, the patient being unable to explain what it is he dreads. Sometimes there is intense anxiety and sometimes a feeling of guilt. Often such stimulation causes stronger and purer emotion than occurs in real life. Whereas it is the nature of the human situation that feelings of delight and joy come more rarely than feelings of misery, an ecstatic feeling that all problems are soluble can be brought about by electrical stimulation of parts of the temporal lobe.[4]

Although this excerpt does not cover *all* emotions, it strongly supports the idea that the brain is "wired" for all emotions and that all emotions can be produced by the appropriate neural inputs. Consequently, the essential requirement for producing an emotion is the brain's detection of a situation in which the emotion is appropriate. The brain then triggers the emotion by generating neural pulses that are the natural equivalent of the electrical stimulation mentioned above.

Notice that we are wired for ecstasy and for guilt. In a sense, then, we are wired for some of the most uncommonly delightful and some of the most ordinary and punishing effects of religion. I can't refrain from noting, however, that this is not the kind of observation that would appear in an "Are We Wired for Religion?" article in the newpaper's religion section.

We are also wired for pleasure and for pain. Pain will be discussed later. As with emotions, pleasure can be produced by electrical stimulation of certain sites in the brain. Animals also have these sites. An electrode can be implanted in one of these so-called *pleasure centers* in an animal's brain, and things can be arranged so the animal can stimulate its own pleasure center by pressing a lever. When this is done to a rat, for example, it will constantly press the lever for hours, ignoring food or anything else.

In humans, an electrode placed in the *septal area* of the brain's *limbic system* can produce pleasant feelings, somewhat like sexual pleasure. In studies by R. G. Heath on conscious, but psychotic patients, electrical stimulation of the septal area produced the following changes:

> Expressions of anguish and despair changed precipitously to expressions of optimism and elaborations of pleasant experiences, past and anticipated. Patients could calculate more quickly than before stimulation. Memory and recall were enhanced. One patient on the verge of tears described himself as somehow responsible for his father's near-fatal illness. When the septal region was stimulated, he immediately terminated this conversation and within fifteen seconds exhibited a broad grin as he discussed plans to date and seduce a girl friend. When asked why he had changed the conversation so abruptly, he replied that the plans concerning the girl had suddenly come to him. . . .
>
> Another patient, an epileptic, was one day agitated, violent and psychotic. The septal region was then stimulated without the patient knowing it. Almost instantly his behavioural state changed from one of disorganization, rage and persecution to one of happiness and mild euphoria. He described the beginning of a sexual motive state.[5]

The artificial electrical stimulation produced an effect similar to that of natural neural and biochemical stimulation of this part of the brain, which is involved in initiating sexual activity and also with producing the pleasure associated with sex. The patients themselves could not explain why their thinking had changed from extreme unhappiness to happy thoughts related to sex.

THE EVOLUTIONARY FUNCTION OF FEELINGS

The nervous system controls a person's behavior. In chapter 8, we saw that nervous systems evolved to promote the biological fitness of animals. Nervous systems generate behavior that makes survival and reproduction more likely.

In humans, and probably in some other animals, both conscious and subconscious processes are involved in controlling behavior. Relatively simple problems concerned with survival, such as those involving the internal maintenance of the body, can be handled routinely by subconscious neural mechanisms. Some examples are the control of digestive processes, the heartbeat, and the body's temperature. Such processes are handled by certain parts of the brain located above the spinal cord and below the cerebral cortex. Antonio Damasio identifies these parts that control the *internal milieu* as the *limbic system*, the *hypothalamus*, and the *brainstem*.[6] Gerald Edelman labels these brain regions the *limbic-brainstem system*.[7] One does not have direct awareness of the operation of these systems. For example, you may become aware of a quickening of your heartbeat when you notice it is beating more rapidly, but you are not aware of the brain's decision to increase your heart rate.

As with a thermostat in a home heating system, the limbic-brainstem system has goals, such as keeping body temperature within a certain range. That the system is successful in achieving its goal is a result of its efficient design, evolved during thousands of generations of trial and error. For such subconscious brain processes, the words "goal" and "design" are metaphors, since the system does not consciously strive to achieve its goals, and its design originated by impersonal evolutionary processes. Our genes specify the development of this system, constraining it to promote our survival and procreation. We will see later that the limbic-brainstem system plays a large role in the functioning of our emotions.

Conscious parts of a person's nervous system also serve the evolutionary needs of the organism. The world outside our bodies generates very complicated and novel problems that are most effectively handled by one of evolution's most powerful and flexible products: the conscious brain. It handles such complex issues as deciding what advice to give someone or planning how to drive to an unfamiliar location in heavy traffic.

The regions of the brain that are active when conscious activity occurs include the *cerebral cortex*, covering the outer part of most of the brain, and the *thalamus*, located in the middle of the brain between the two cerebral hemispheres. The thalamus has extremely large numbers of connections to all parts of the cerebral cortex. The combination of the cerebral cortex and the thalamus is called the *thalamocortical system*. It is involved in all aspects of what we normally call *mental activity*. It interprets sense data and infers what is going on around a person. It is involved in the process of forming and recalling memories. It makes our decisions, composes our speech, and plans our actions. It is essential for consciousness.

Evolution has conspired with the subconscious parts of the brain and the rest of the body to insure that the thinking and decision-making processes of the conscious brain are compelled to serve the organism's need for survival and reproduction. The interacting conscious and subconscious systems are so strongly interconnected that even one's supposedly "purely rational choices" are strongly biased by motivations arising in more primitive and subconscious portions of the brain. These motivations tend to steer one toward actions that enhance one's biological fitness.

These motivations include such feelings and emotions as pleasure, pain, fear, anger, hunger, disgust, and happiness. Pleasure leads us to do something that increases the likelihood of our survival or reproduction. Pain leads us to protect our bodies from harm. Fear leads us to avoid dangerous situations, and anger prepares us to fight. Hunger leads us to eat, and disgust prevents us from eating spoiled or contaminated food. Happiness seems to be a more complicated feeling. Maybe it is a combination of relaxation and intellectual pleasure produced when things which one values are happening or when one's circumstances seem to be satisfactory or improving. More biologically, it may be a sort of "all's well" state that allows such internal processes as heart rate and the activity of the digestive system to be set to their most healthful values when there are no urgent external demands on the body. It may also involve some low-level stimulation of some pleasure center in the brain.

It is evident that our feelings serve our biological needs. It is not so clear how the body and the brain generate our feelings. Even more of a puzzle is why our feelings *seem the way they do to us.*

Can a physical system produce these subjective effects? We will
tackle these issues in the following sections.

VALUES, VALUES, EVERYWHERE

Let's consider how mental processes trigger our feelings. Some feelings, such as fear or anger, often start with some conscious activity.
We learn that a disturbing situation is developing, or we think of a disturbing possibility. In such cases, consciousness is involved at the
start of the emotional process. How does the mind conclude that
some news, situation, or thought is *disturbing*? No one knows exactly,
but I will suggest how this may happen, based largely on Antonio
Damasio's *Descartes' Error*. He proposes specific roles for the thalamocortical system, the limbic-brainstem system, and other parts of the
body in the operation of human emotions and motivations.

Innate values are part of our genetic heritage. They are *instinctive
appraisals* of certain situations. A newborn infant has few previous
experiences of things in the world outside the womb. Even at birth,
however, the infant cries when painful things happen. He or she seems
to find eating pleasurable from the start. Innate values cause infants to
avoid pain and enjoy eating. People also seem to have built-in tendencies to fear snakes and great heights. Individuals have varying
degrees of success in controlling and overcoming these innate fears.

Norton Nelkin has described studies that show that few-month-
old babies are delighted if they can control the movement of
mobiles suspended above them.[8] The experiment involved special
riggings that allowed the babies to move the mobiles by moving
their heads. What is most interesting is that other groups of infants
who could not move the mobiles, or for whom the mobiles moved
outside their control, did not take delight in the mobiles or their
motions. The studies suggest that *we have an inborn desire to control things*. Apparently this is an innate value. Nelkin argues that
the basic idea of being in control plays a large role in allowing the
infant to separate the world into *self* and *nonself* and in developing
the idea of the *will*.

As consciousness develops, an infant becomes aware of the attractive or repulsive effects of certain situations and *begins to associate positive or negative values with these situations*. The innate values lead, through a person's experiences, to the automatic assignment of *attached values* to all things and situations. Our experiences produce links of certain innate values with everything that we observe in the outside world.

An infant learns to associate its mother with pleasurable experiences and protection from painful or frightening situations. A child learns to appreciate objects that it can manipulate, since this is inherently rewarding. In the process, the child learns to value his or her toys. Following a painful experience with a dog, a child may fear dogs, with a negative value attached to dogs in his or her mind. By such processes, innate pleasure, pain, or fear leads to the attachment of positive or negative value to things or situations associated with the experiences. This process of attaching values to things and situations continues throughout a person's life. In the end, everything has values associated with it. Among other things, these *attached values* allow us to make choices in supermarkets.

A thing's attached values are immediately perceived when one thinks about the thing. Think of a puppy, a kitten, AIDS, and death. We can't think of these things without thinking about their attached values. This is also true for people or situations such as Abraham Lincoln, Adolf Hitler, Jesus, crucifixion, walking on coals, taking examinations in school, going to the dentist, eating cake with ice cream, and so on. Values are a major part of the meaning of words and concepts stored in our memories.

The values we attach to words or categories of things or situations carry more information than just positive or negative attitudes. For example, think of the words "rattlesnake" and "baby-killer." Both words have negative connotations, but "rattlesnake" typically stimulates *fear*, while "baby-killer" provokes *anger*, *hatred*, or *contempt*. So it seems there are a number of dimensions to attached values. "Rattlesnake" is usually high in the fear dimension, while "baby-killer" ranks high on certain other emotions.

The preceding discussion treats *values* as rather "mental" or abstract entities. We will see later, however, that values are the capability of certain neural circuits to generate signals which steer our decisions toward actions that enhance our biological fitness. The signals are our *motivations*. When we are conscious of these signals, they become our *feelings. Innate values* arise in circuitry that is specified by our genes. *Attached values* arise in circuitry that has been "trained" by our experiences to produce value signals that are aligned with our innate values.

TRIGGERING AND ACTIVATING THE EMOTIONS

Sometimes our emotions are triggered by innate mechanisms, as discussed above. The innate values that trigger such emotions are "wired into us" and were specified by our genes. An infant's contentment or happiness while being fed is an example. Antonio Damasio calls such emotions *primary emotions*.[9] These emotions are probably triggered in the limbic-brainstem system. They occur when a situation fits a genetically specified pattern that is already present, or programmed, at birth.

Frequently, however, our emotions are triggered by conscious thoughts in the thalamocortical system. Damasio calls these *secondary emotions*.[10] He maintains that the prefrontal region of the cerebral cortex is deeply involved in the thoughts that trigger emotions. The thoughts have attached values, or emotional baggage, because of our previous experiences.

All emotions, whether *triggered* by innate, subconscious mechanisms or by value-laden conscious thoughts, seem to be *activated* by stimulation of parts of the limbic-brainstem system. *That is, emotions triggered by conscious thoughts are apparently activated by the same mechanisms that express emotions generated by innate, subconscious triggers.*

The *amygdala* is deeply involved in activating both the primary and the secondary emotions. It is a small, almond-shaped organ found on both sides of the brain within the limbic system. By connections through the thalamus, it receives signals from the senses by which it can trigger primary emotions. Neurons within the amygdala can respond to such specific and complex sensory patterns as visually observed faces.[11] By connections from the cerebral cortex, the amygdala can receive signals that trigger the secondary emotions. It is interconnected with the hypothalamus and the *brainstem reticular formation*, through which it can control much of the expression of emotions.

Animal experiments have shown that behavior that we associate with fear and anger can be eliminated by removing the amygdala or its connections with other parts of the brain. Damasio mentions the case of a woman in whom the amygdala is not functional on either side of the brain, but the rest of the brain is intact.[12] The woman is afflicted with a "life-long pattern of personal and social inadequacy. There is no doubt that the range and appropriateness of her emotions are impaired and that she has little concern for the problematic situations in which she gets herself."

The excitation of the amygdala and another part of the limbic system, *the anterior cingulate*, leads to emotional arousal. Damasio describes four means by which emotional arousal occurs:[13]

- Signals from the brainstem reticular formation signal the *autonomic, or involuntary, nervous system*. This can produce many physical effects, such as increased heart rate, sweating, dilated pupils, and the cessation of digestive processes.
- Neural signals also activate motor neurons that control certain muscles, producing the changes in body posture and facial expressions that are part of the "body language" of the emotions.
- In a slower process, the *endocrine glands* emit certain biochemicals, such as *adrenaline*, that change the state of the brain and the body as they spread in the bloodstream.
- By another process involving chemical messages, the brainstem and certain clusters of neurons known as the *basal forebrain nuclei* change the state of other parts of the brain, including the cerebral cortex.

The list suggests ways that we may actually *feel* our emotions. We become aware of the excitation of the autonomic nervous system through such effects as a racing heart, profuse sweating, hairs standing on end, and various "gut feelings." These effects are communicated to the conscious brain by sensory neurons. This is obviously an important way by which we feel emotions. One speaks, for example, of "heartfelt love" because the effects of love are sometimes made evident by an increased pulse rate.

The *chemicals* emitted by the endocrine glands and the basal forebrain nuclei also affect our consciousness. We frequently speak of the "rush" produced by a "burst of adrenaline." One's consciousness is altered by the brain's changing chemical environment in the presence of strong emotions. Perhaps the chemical changes cause the emotional effects on thinking processes described by Damasio. For positive emotions, the changes are that "the generation of images is rapid, their diversity wide, and reasoning may be fast though not necessarily efficient," while for negative emotions, "the generation of images is slow, their diversity small, and reasoning inefficient."[14] Severe negative emotions may lead a person into a downward spiral of *depression* in which one is not only depressed, but unable to think clearly about improving one's situation.

Besides the effects of physical and chemical changes in the body and the brain, emotions may affect one's consciousness by direct neural signals from the subconscious limbic-brainstem system back to the conscious thalamocortical system. Such effects could operate in a loop if the emotional stimulus originating in the conscious brain triggered the limbic-brainstem emotion generators and sent a strong signal back to the conscious brain.

The conscious brain, in thinking about value-laden concepts, is aware of the emotion-causing effects of these ideas. Sometimes one may experience these emotions in a rather rational manner, even without the physical or chemical effects described above. The human brain is capable of doing this since we sometimes know that certain comments will produce strong emotional effects *in other people*, even when we react only weakly to the comments.

By now it should be clear that there are plenty of ways by

which we may feel our own emotions, although many of the processes are not well understood. There does not seem to be a need for any supernatural explanations of the emotions. They seem to present some *puzzles* to be solved, but no *mysteries* that will defy all possible explanations.

PAINS

Our feelings of pain and pleasure suggest some interesting questions about the mind/body relationship. Why do we have pains? Why do psychological effects play a large role in how we feel pain? Although we accept this as natural, just how is it that we feel that pain is bad and pleasure is good? Is it built into our nervous systems? How do streams of pulses carried by neurons become pains and pleasures in the brain? This and the following sections will discuss such questions.

Pain has survival value. It helps us avoid doing things that damage the body. A few people have a congenital condition that causes them to be completely insensitive to pain. They are often happy and giggly even though they are damaging their joints by bending them too far, and they frequently cut or burn themselves severely. These unfortunate people tend to die young because of accumulated damage, injuries, or accidents. Clearly, pain helps us take care of our bodies.

Pain is somewhat like touch in that we are conscious of at least the rough location of a pain. Unlike touch, it carries little information about *what* is in contact with the skin. It does not inform us directly about what is damaging the body. Instead, pain is only a very unpleasant feeling that appears to come from a specific part of the body. It is a sort of built-in value judgment that *something bad* is happening in the indicated part of the body. It seems *built-in* because the judgment appears even if we are not conscious of the body's problem. We sometimes know what causes the pain, but this is based on learning from previous experiences.

Because people will pay to reduce pain, considerable work has been done to learn how to control it. Much has been learned, particularly about the portions of the pain process that occur outside the brain's conscious regions.

Pain is not just very intense feelings associated with the sense of touch. The neurons that detect and transmit pain signals are separate and of a different type from those sensitive to touch. Pain signals start in tiny sensors called *nociceptors* at the ends of pain-sensing neurons. Under unusual mechanical or thermal conditions, nociceptors produce pulses in their attached neurons. There are two kinds of pain neurons, *Aδ fibers* and *C fibers*. Aδ fibers carry signals associated with the tingling to sharp pains that are perceived first when a painful situation suddenly develops. C fibers transmit signals of duller, longer-lasting pain sensations.[15] Pains can be produced artificially by electrical stimulation of Aδ and C fibers.

Although we tend to think of pain as a straightforward result of having pulses within certain neurons, what happens in the later stages of the pain process can reduce or heighten one's conscious perception of pain. Pain-sensitive fibers connect to other neurons at junctions in the upper spinal cord and brainstem, along neural pathways to conscious areas of the brain. The *gate control theory of pain*, proposed by Ronald Melzack and Patrick D. Wall in 1965, was important in stimulating studies of how pain is controlled at these junctions. These studies led to the discovery of *endogenous opioids*, chemicals produced within the body that reduce pain by blocking the signals at these neural junctions. These natural pain-blockers are chemically related to derivatives of opium long used to reduce pain.

Neural connections from the upper parts of the brain control pain by sending signals down to the neural junctions, where opioids are apparently released, reducing the pain signals sent upward. By this means, psychological processes can have a strong effect on the sensation of pain.

There are dramatic examples of the control of pain by psychological effects. In World War II, a systematic study of severely wounded soldiers found that they often experienced little pain. Many of the soldiers themselves were surprised at their lack of pain. Henry Beecher, an anesthesiologist who was involved in this study, "concluded that the perception of pain depends very much on its context. The pain of an injured soldier on the battlefield would presumably be mitigated by the imagined benefits of being removed from danger, whereas a similar injury in a domestic setting might raise quite a different set of thoughts that could exacerbate the pain (loss of work, financial liability, and so on)."[16]

Another example of psychological effects on perceived pain occurs when a doctor prescribes a normally ineffective and harmless drug or *placebo* to reduce pain. Under these conditions, patients usually experience some improvement. In hospitals, "Typically, 3 out of 4 patients suffering from postoperative wound pain report satisfactory pain relief after an injection of sterile saline. The researchers who carried out this study noted that the responders were indistinguishable from the nonresponders, both in the apparent severity of their pain and in the nature of their characters."[17]

Although it is psychological in origin, the *placebo effect* involves real changes in patients' bodies. Those whose pain is reduced by a placebo have increased levels of endogenous opioids in their bloodsteams.[18] These natural painkillers are produced and injected into the circulatory system by endocrine organs under control of the brain. A drug called *naloxone* can block the effects of opioids. As a test, it was administered to some of the postoperative patients mentioned above. The placebo effect disappeared, supporting the hypothesis that natural opioids produce the effect.

It often surprises people that pains produced by real physical effects in the body can be influenced by psychological effects. After all, don't our mental processes occur outside the material world? (Well, no, they don't.) The feeling of surprise betrays a tendency to think of the mind as a transcendental entity. It comes from making too great a distinction between the world of matter and the "world" of conscious experiences.

Let's look at psychological effects on pain from another angle. Pain, and the psychological events that reduce it, result from electrochemical events in the nervous system. Why shouldn't the one kind of event influence the other? Considering how thoroughly the brain's neurons are interconnected, it is not surprising that such effects exist.

PRIMITIVE MOTIVATIONS IN MODERN MINDS

Once upon a time, there was no consciousness on Earth. There were animals, but they were aware of nothing. Their actions were instinctive responses. As time went by, they became capable of improving their responses by learning from their experiences. When the capabilities of their brains reached an advanced stage, they began to combine information coming from their various senses and they constructed mental images of what was happening in the world outside their bodies. Gradually, consciousness came into being.

Imagine a life-form that was just beginning to develop consciousness. The new, conscious part of its nervous system can be called the *New System*. Like its ancestors, the organism had instinctive behaviors that allowed it to respond automatically to developments in its surroundings. The part of its nervous system that responded in such an automatic way can be called the *Old System*. The New System, or NS, could use its sensory data and knowledge from previous experiences to form a mental picture of its current situation. Then it could evaluate various options and make *informed decisions* about what actions to take. The NS was very flexible in matching its responses to the perceived situation, while the Old System (OS) could only trigger one or more of its built-in responses. Eventually, the NS took control of activities that require flexible responses to complex situations. In practice, this meant that many actions involving the external world were brought under control of the NS, while the OS kept control of routine functions, such as breathing, that could be handled by standard responses operating outside of consciousness.

With the NS's extra flexibility, the organism increased its evolutionary fitness by making choices that were more appropriate for specific situations. To accomplish this, however, the conscious organism needed *goals and values* to help it select options that would enhance its chances of surviving and reproducing. It could have learned these values by trial and error, but this would have resulted in most individuals making some bad choices and dying prematurely. Fortunately, there was a better way to set up appropriate goals and values in the NS. This was done by transferring the values of the OS to the NS. During previous eons, the OS had gone through a painful trial-and-error design process. As a result, the OS had developed a suitable repertoire of standard responses that helped the organism's ancestors survive. The practical wisdom embodied in the OS was passed on to the NS through the transfer of *value signals*.

The neurons that triggered the OS standard responses sent their value signals to the NS, to tell it what to do. The NS, however, did not use these signals in the same way as the OS. That is, the signals were not used to trigger standard responses. Instead, in the NS, the OS signals were compared with other signals generated by the NS, and a decision would be made. If no strong NS signals were present, the outcome might be similar to a standard response of the OS. If the NS signals were stronger than the ones from the OS, then the decision produced a response based more on the NS's mental processes than on the traditional OS values.

When the NS made a decision between a number of imagined choices, how did it *evaluate* the various options? That is, where did it get the criteria it used to judge the relative merit of the different possibilities? Options based on standard OS responses were evaluated using the value signals sent from the OS. Options suggested by the NS were assigned merit according to their *innate or attached values*. Directly or indirectly, all the value signals were produced by the preexisting mechanisms for generating values in the OS. Thus, the OS values played extremely important roles in controlling the conscious decisions of the NS. Even the subconscious process by which *attention* selected the most important information streaming from the outside world needed criteria for evaluating what is important. This evaluation was ultimately based on the OS values.

Even today our brains probably operate under something like this scenario. The OS is the limbic-brainstem system; the NS is the thalamocortical system. Our (NS) processes of planning and decision making are heavily dependent on the OS for the weights that are assigned to the different options. Because of this, our thinking is strongly guided by OS values that have the goal of increasing our evolutionary fitness.

In our planning and decision making, we are constantly aware of what we like and what we want to do. These "likes" and "wants" may be hard to explain because they do not arise from verbal, or even rational, thought. They are the direct or indirect influences of the OS on the operation of the NS. For example, we find that we like sweets, but we can't explain why. If asked, we may reply, "I just like them, doesn't everyone?" Our liking them is not due to rational considerations but to the fact that we are programmed by our genes to obey the evolutionary goals of survival and procreation. Eating sweets serves the goal of consuming calories needed to provide the energy needed by our bodies.

We are brainy animals, and there are sometimes very indirect connections between our personal goals and the goals of evolutionary fitness. As an example, we may think very rationally about which college to attend. The thought processes are highly conscious and of such a high level that they seem to be going on entirely within the NS. If we examine how the decisions are made, however, the different options may be evaluated by considering likely long-term goals of earnings, prestige, or doing interesting work, and such short-term goals as enjoying college, staying near one's family, being with friends, having a rich social life, and so on. The goals of social and material success are motivated by our biological drives. Even the desire for interesting work is based upon the innate value that drives infants and children to want to learn. Ultimately, these motives are traceable to OS values promoting survival and reproduction.

Sometimes we obey evolutionary drives until they no longer enhance our fitness. We may eat too much. A less obvious example is our search for eternal salvation. Our natural goal of survival leads us to seek immortality. The Bible professes to tell us how to attain eternal life. So, the possibility of using faith and/or good works to be saved seems highly attractive since it appeals to our basic animal goal of survival. By humbly keeping the faith and keeping our eyes on the prize, we may be able to accept the Bible's promise in spite of its implausibility. Some have suffered martyrdom while following this promise. Clearly, we can subvert evolutionary drives, so they fail to enhance our biological fitness. Some may argue that martyrdom serves a social purpose because it supports the social group in the name of which the martyr dies. Does this enhance the biological fitness of the human species? If the martyr is a misguided member of a cult that encourages false hopes of immortality, the martyrdom is clearly tragic. It serves the cult that promotes martyrdom, while destroying the individual. In such a case it is not evident that the martyrdom enhances the biological fitness of the human species.

Our feelings, motivations, and desires issue from our evolutionary origins. From this perspective, "getting in touch with our feelings" amounts to paying closer attention to our biological nature. We should not ignore this. To quote Joseph LeDoux, "Many emotions are products of evolutionary wisdom, which probably has more intelligence than all human minds together."[19]

WHY PAIN FEELS BAD AND PLEASURE FEELS GOOD

Why do we automatically assess pleasures as good and pains as bad? Both pains and pleasures start with neural pulses coming from outside the brain. How does our consciousness label them as good or bad? From the evolutionary perspective, of course, pains and pleasures inform us that some situations are to be avoided and others are to be sought. Our question, however, is how *we come to have subjective feelings that are aligned with the values favored by natural selection.*

Consider how a primitive animal which lacks consciousness deals with situations that involve pain and pleasure in conscious animals. In such an animal, the OS triggers various instinctive behaviors. For example, if a part of a snail is damaged, which would produce pain in an animal with consciousness, the snail withdraws its body into its shell to protect itself. Assuming that snails are not conscious, this is an automatic response. Insects mate rather automatically when certain conditions are satisfied. Presumably they are not led to intercourse by pleasurable foreplay. Even relatively complicated behaviors can be produced without pain, pleasure, or other conscious motivations. Communications between foraging honeybees may be an example of this.

In an animal with conscious decision making, however, the responses are not automatic. They are selected according to their perceived merit. A subconscious subsystem of the organism generates *value signals* for different possible responses. The signals are values or weights assigned to all the considered choices. As discussed earlier, the value signals are ultimately based on genetically specified innate values. To be effective in promoting the animal's fitness, the value signals must correspond, at least roughly, to the relative *fitness values* of the various choices. Over the long run, natural selection keeps the value signals close to the levels that maximize fitness.

A conscious human is at least partly aware of the value signals for various potential choices. Value signals are neural signals in the conscious brain, so it is *plausible* that we should be aware of them. Since they are crucial in the decision process, it is *useful* to be aware of them so we can predict our own future behavior. This is important for realistic planning of our actions. By now it may be obvious that *the value signals of which we are conscious are our feelings*. They are our conscious motivations, wants, needs, pains, and pleasures.

In a person's awareness *a value signal seems like a sort of mental force that tends to control the decision-making process*. Our actions are ultimately controlled by our decisions, which, in turn, are controlled by our value signals. Since so much of the process of generating the brain's values is subconscious, these values, or feelings, may seem to be superimposed on our thinking from the outside. They may seem to be things that happen to us. For example, when a person is dismayed by her less than admirable motivations, she may blame them on an outside cause, as in, "The Devil made me do it."

On the other hand, we usually accept our values (feelings) as part of our conscious selves. This appears reasonable since it seems to us that such feelings were always present in us in the past, and they have always been very important in determining how we behave. Thus, we usually say, "I feel . . . ," not, "I find that I feel . . ."

The preceding discussion links neural signals generated by subconscious brain processes with feelings experienced by a conscious person. The word "value" was especially useful since it can describe: (1) an indicator, on the evolutionary fitness scale, of the relative worth of a proposed action; (2) the strength of a neural signal that influences a mental decision process; and (3) the consciously experienced intrinsic worth of a potential decision choice. All of these are useful in thinking about feelings.

The feeling of pleasure is probably produced by a strong positive value signal sent by the OS to the NS. It is "positive" in that it tends to support a certain situation or activity. Such a value signal makes a considered option *attractive* to consider and *reinforces* any decision made in favor of maintaining a pleasurable situation or continuing a pleasurable activity.

From the neurophysical perspective, the "attractive" nature of a situation, activity, or option arises because a strong value signal makes a situation or activity more likely to be maintained and the option more likely to be chosen. From the subjective perspective, the essence of the attractive nature is that one is spontaneously inclined to decide in favor of maintaining the situation or activity and deciding in favor of the supported option. A decision to choose the favored option is reinforced by the strengthening that occurs in the "decided" feeling if the pleasant feelings continue.

Besides the positive value signal, the pleasure process has another dimension, which associates the pleasure with a particular location in the body. This makes the signal more specific, helping a person know *which activity* the OS is encouraging. Similarly, for pain, the fact that it is localized helps one learn what to do to reduce it.

Pain is probably produced by strong negative value signals sent by the OS to the NS. Of course, the ultimate cause of these signals was pain signals from some of the body's nociceptors. Pain is "negative" in the sense that it *opposes* some aspect of the body's current situation. Because it is intense, disruptive, and arousing, it tends to interrupt rewarding or goal-seeking behavior. It allows one no peace or pleasure, and causes a person to want to deal with it. Very early in life, we adopt the attitude that pain is to be avoided. We come to feel that pain comes from a threat to our normal functioning or survival. We regard pain as opposed to the self, aligning our feelings with the evolutionary values of the OS.

As far as I know, no one has ever pointed to the axon of a neuron and announced, "This axon carries a value signal from . . ." As a plausible hypothesis, however, the idea of a value signal helps us imagine how emotions and motivations affect our decisions. The idea serves as a bridge between the world of neurons and the world of our subjective feelings.

WHO OR WHAT IS IN CONTROL?

The picture of motives and emotions painted here suggests a particular view of human nature. It implies that it is misleading to claim we are basically rational. Although we have a much higher ability to reason than other animals, reasoning usually does not *control* our everyday decisions. Typically, we use reasoning as a tool to help us decide whether a proposed course of action would achieve *what we want it to do*. The basic *goals* do not result from the reasoning process. What really counts is whether potential choices are in harmony with our basic evolutionary values. Although our complicated thought processes may hide our fundamental motivations, subconscious mechanisms inherited from our animal ancestors motivate our behavior and determine our decisions.

The conclusion that *mechanisms within the brain determine our choices* indicates a conflict between the picture of decision making given here and some common ideas about *free will*. This will be discussed in the next two chapters. We will find that the views expressed here *do* contradict some traditional free-will beliefs.

NOTES

1. Antonio R. Damasio, *Descartes' Error* (New York: Grosset/Putnam, 1994).

2. Gerald M. Edelman, *Bright Air, Brilliant Fire: On the Matter of the Mind* (New York: BasicBooks, 1992).

3. Joseph LeDoux, *The Emotional Brain: The Mysterious Underpinnings of Emotional Life* (New York: Simon & Schuster, Touchstone Books, 1996).

4. Peter W. Nathan, "Nervous System," in *The Oxford Companion to the Mind*, ed. Richard L. Gregory (Oxford: Oxford University Press, 1987), p. 527.

5. R. G. Heath, quoted by Nathan, "Nervous System," p. 529.

6. Damasio, *Descartes' Error*, p. 118.

7. Edelman, *Bright Air, Brilliant Fire*, p. 117.

8. Norton Nelkin, *Consciousness and the Origins of Thought* (Cambridge: Cambridge University Press, 1996), pp. 301, 302.

9. Damasio, *Descartes' Error*, pp. 131–34.

10. Ibid., pp. 134–39.

11. Dale Purves et al., eds., *Neuroscience* (Sunderland, Mass.: Sinauer Associates, 1997), p. 520.

12. Damasio, *Descartes' Error*, p. 69.

13. Ibid., p. 138.

14. Ibid., p. 147.

15. Purves et al., *Neuroscience*, p. 166.

16. Ibid., p. 173.

17. Ibid., p. 174.

18. O. T. Phillipson, *The Oxford Companion to the Mind*, p. 222.

19. LeDoux, *The Emotional Brain*, p. 36.

13

PHYSICS, THE QUANTUM, AND FREE WILL

I N DISCUSSING *FREE WILL*, I will focus on a dramatic conflict between two opposing views of reality. Both views contribute to our everyday beliefs. One view emphasizes the presence of supernatural occurrences in our world. It implies that our acts are essentially unpredictable. The other view asserts that the world is orderly and operates by natural processes. This side maintains that our acts are predictable in principle, if not in practice.

On the one hand, a certain form of free will is often held for religious and traditional reasons. This view is present in much of Christianity, Judaism, Islam, and Zoroastrianism. It maintains that when we do something wrong, *we could have selected a better choice and carried it out under exactly the same conditions*. As we shall see, this view implies that we have supernatural powers that we can use to intervene and control the outcome when we make moral decisions. As will be explained below, I will call this the belief in *Objective Free Will*.

Opposed to Objective Free Will is the belief that the world operates by purely natural forces. In this view, everything functions according to scientific laws, and everything happens *the only way it can*. From this perspective, Objective Free Will seems unlikely since it would introduce anarchy into what is otherwise a very orderly world. Such a view can be traced back to Democritus, who was mentioned in chapter 5 in discussing the origin of the idea of atoms. This view maintains that *our decisions are determined at the most fundamental level by the detailed physical circumstances*. In reality we could not have done otherwise. Such a view, which denies free will and maintains that everything happens the only way it *can* happen, is called *determinism*.

In this and the following chapters, it will be shown that it is reasonable to believe that: (1) *The choices we make are the only choices we could have made without violating the laws of nature,* and (2) *We have control over our own choices.* Although these claims may seem contradictory, they are not. The conclusion that will be reached on the free-will issue is not an extreme position, but it will suggest that we need to change many of our beliefs.

OBJECTIVE FREE WILL: THE TRADITIONAL BELIEF

The traditional idea of free will is what I have called *Objective Free Will*. I will represent it by the acronym **OFW**. The fancy script emphasizes its ancient and hallowed origins and its connection with supernatural ideas. When a person makes a decision, according to **OFW**, there is a real possibility that a different decision could have been made *under exactly the same circumstances*. For many Christians, this ability to choose is one of the powers of the human soul. I will argue against **OFW**.

Why did I call this kind of free will *Objective Free Will*? I will argue below that only one result of a decision is possible, given the entire situation and the orderliness of physical processes. If **OFW** is true, however, a person has the ability *to violate this orderliness and cause a different outcome to occur*. This would be due to something outside the natural: a *supernatural* effect. It would be *a minor miracle*. It would introduce a sort of capriciousness or arbitrariness into the physical world.

OFW is consistent with many of our beliefs and intuitions. We feel that we really do choose between two or more possibilities each time we make a decision. In disagreeing with **OFW**, I do not claim that this feeling is wrong, but that the process of deciding *selected the only choice that could have been made under the exact circumstances*. We are not completely aware of the processes involved in reaching a decision. These processes are orderly and they lead to the only natural outcome. *In the next chapter we will see that a different kind of free will that I support also agrees with our intuitive feelings about freely choosing a decision's outcome.*

If **OFW** is true, there is a profound sense in which a person can *originate* an action. In this picture, the earliest cause of a particular action would appear at the time of the person's *act of the will*. Before the act of the will, there was no *physical reason* why this particular action should result. The person, through the supernatural power that we can call *origination*, can cause something to happen that was not the natural outcome of the detailed physical situation.

If **OFW** is true, things can happen when people make decisions that change the *natural* outcome into a different, *chosen*, outcome. *Such decisions would produce changes in the real world.* In principle, an experiment could detect such a minor miracle as the operation of neurons in the brain in a way that is normally impossible. This would cause a person to perform a different act than *should have happened*, according to the laws of nature. This minor miracle could be detected. Thus, **OFW** implies that minor miracles occur that are, *at least in principle*, detectable. **OFW** implies that a person's free will produces real, *objectively detectable* changes in the world. It is therefore appropriate to call it *Objective Free Will*.

THE ORIGINS OF OBJECTIVE FREE WILL

It is easy to see why many Christians, Jews, Muslims, and Zoroastrians share a belief in a strong form of free will such as **OFW**. According to these groups, God is just and good but rewards or punishes us for our actions. If God punishes or rewards us for our deeds, it is logical to believe that we normally have a real ability *to choose good or evil, in spite of the detailed circumstances that apply when we make our choices*. Otherwise, God would seem unjust to punish or reward a person for making a choice that is merely the unavoidable result of the operation of natural forces.

In Christian and related faiths, **OFW** is a power given to us by God. God is also believed to have created the entire universe by an act of will. In these religions, the human will is a limited version of the all-powerful divine will, which is superior to all physical laws and can cause them to be violated. Thus, this kind of will (**OFW**) must have a supernatural nature, if it exists.

The classical Greek culture, which had so much influence on the entire Roman Empire, had some ideas that were not consistent with **OFW**. Many Greeks believed that a person could make only that choice that was favored by one's particular intellectual and emotional constitution. To them, outstanding behavior originated from unusual knowledge rather than from unusual exertions of the will. Many Greeks and non-Greeks also believed in *the inevitability of fate*, which is contrary to **OFW**.

Albrecht Dihle, a German scholar concerned with philosophy, religion, and classical studies, noted that it was "the common creed of Greek philosophy that nature is strictly ordered by reason and does not admit exceptions to its rules."[1] The Greeks held a strong belief in the orderliness of all processes in the physical world. Consequently, they would probably have been reluctant to accept the arbitrariness of human choices that is implied by **OFW**. According to one powerful school of Greek philosophy, the *Stoics*, "Even the most insignificant detail in everyday life has been predetermined by nature or fate in the very same way as cosmic phenomena."[2] This leaves no freedom of the type that is needed for **OFW**.

Saint Paul, often called Christianity's first theologian, did not mention free will. Paul did, however, maintain that humans are free and responsible for their moral conduct. This suggests that the later Christian idea of **OFW** was consistent with the beliefs of the earlier Christians.

Paul also emphasized the doctrine of *predestination*, as in Rom. 8:29–30. According to this doctrine, one's salvation depends on whether one is chosen by God. This doctrine tends to contradict **OFW**, although carefully constructed theological interpretations can avoid a contradiction. Within Christianity, supporters of free will and predestination have been involved in serious disputes. Islam has also had problems in reconciling the two ideas.[3]

About two centuries after the life of Paul, the influential non-Christian and Neoplatonist philosopher Plotinus (204–270 C.E.) taught that free will results from a *physically uncaused act of a spiritual soul*. The following quote is an explicit statement in support of **OFW**. Concerning our free acts, he wrote:

> But when our Soul holds to its Reason-Principle, to the guide, pure and detached and native to itself, only then can we speak of personal operation, of voluntary act. Things so done may truly be described as our doing, for they have no other source. They are the issue of the unmingled Soul. . . .
>
> All things and events are foreshown and brought into being by causes; but the causation is of two kinds; there are results originating from the Soul and results due to other causes, those of the environment.[4]

Saint Augustine of Hippo (354–430 C.E.) was a major developer of the **OFW** concept in Christian theology.[5] In his book *On Free Choice of the Will* he makes the argument that was given at the start of this section:

> Each evil man is the cause of his own evildoing . . . evil deeds are punished by the justice of God. It would not be just to punish evil deeds if they were not done willfully.[6]

Augustine discussed many questions such as how evil can exist in a world created by a good and all-powerful God. He concluded that evil done by us or by corrupted angels comes from the misuse of **OFW** that was given to the angels and us by God. The notion of free will that seems to be implied in the Bible became an important explicit concept in Christian theological discussions after Augustine.

Although the reformers Calvin and Luther challenged beliefs about **OFW** on religious grounds, it appears that in all Christian denominations in our society many members hold beliefs that make sense only if minor miracles really happen. *The idea of punishment by God makes sense only if people are really able to use minor miracles to shift their moral decisions away from their natural outcomes.* Now, let's consider arguments against **OFW** that are suggested by science.

CLASSICAL PHYSICS AND DETERMINISM

A little more than a century ago, physicists had achieved great successes in understanding the inanimate world. In 1894, so much had been learned by studying forces and the motions of objects, as well as sound, light, heat, electricity, and magnetism, that the prominent American physicist Albert Michelson (1852–1931) proclaimed,

> The more important fundamental laws and facts of physical science have all been discovered, and these are now so firmly established that the possibility of their ever being supplanted in consequence of new discoveries is exceedingly remote. . . . Our future discoveries must be looked for in the sixth place of decimals.[7]

Physics as it was known at about that time is often called *classical physics*. It was soon found to be inaccurate when faced with situations that are far outside our ordinary experiences, such as moving at nearly the speed of light. New discoveries led to new theories. Ironically, one of Michelson's own precise experiments became a major factor leading to one of these new theories: Einstein's *special theory of relativity*. Einstein also produced his *general theory of relativity*. Many people contributed to the new subject areas of *quantum mechanics* and *elementary particles*. The new subfields are sometimes grouped together and called *modern physics*.

To see how classical physics is related to determinism, let's consider the motions of bodies within the Solar System. Before classical physics, during the late Middle Ages, it was believed that moving bodies require a *motive power* to keep them in motion. This was based on the ideas of the ancient Greek philosopher Aristotle. Aristotle thought that the Moon, Sun, and planets required "souls" to maintain their motion.[8] During the Middle Ages, some people thought that angels or the hand of God kept the heavenly objects in motion.

Sir Isaac Newton (1642–1727), an English physicist and mathematician who was one of the world's all-time greatest scientists, found a simple mathematical formula that successfully describes the force of gravity. Using this law along with his laws dealing with the effects of forces on the motion of bodies, Newton explained the motion of bodies within the Solar System. The nearly circular orbits of the planets around the Sun were understood in terms of the Sun's gravitational attraction continually deflecting the direction of the planets' motions. The *sustained movement* of the planets was understood as a natural result of their previous motion and did not need to be maintained by souls, spiritual entities, or anything else.

Although Newton had used natural processes to explain much about the Solar System, he was a devout person and felt that the nearly circular motions of the planets in one direction around the Sun must have been initially set up by God. He also felt that God must intervene occasionally to maintain such regular behavior in the Solar System. Later, Laplace and Lagrange explained how the regular behavior of the Solar System is maintained by natural processes. Laplace made a famous remark to Napoleon that the "hypothesis" of God's existence is not needed to explain the characteristics of the Solar System.[9]

After the work of Laplace and others, the generally accepted scientific belief was that if one knew the *initial conditions* that applied to the Solar System at some earlier time, then the entire future development of the Solar System could be predicted. The initial conditions would describe the exact circumstances at some time by specifying, for example, the precise locations, velocities, and rotational motion of all objects in the Solar System. *The initial conditions combined with the laws of physics would effectively determine all of the future behavior of the system.*

In the eighteenth and nineteenth centuries, scientists learned that all physical phenomena, such as light, sound, heat, electricity, and magnetism, are subject to scientific laws. It was observed that the laws of physics apply to living matter as well as nonliving matter. All of this led to the belief that knowledge of the initial conditions and the scientific laws was all that was needed to allow scientists to explain the future behavior of the universe. Philosophers call this belief *physical determinism.*

According to determinism, the future behavior of any system is completely predictable, in principle, if enough is known about the system. In its predictability and reliability, this idea paints a picture of a very orderly and mechanical universe that resembles the operation of a clock. According to determinism, *everything* evolves in an inevitable way that is precisely determined at the time of the origin of the universe. Since we are part of the natural world, determinism was expected to apply to our behavior as well. This set of beliefs did not allow people the freedom of action that is implied by **OFW**.

If you believe in **OFW**, you may argue that free will is a result of spiritual rather than material causes. If free will is a power of a spiritual soul, it could be exempt from the determinism of the physical world. From this perspective, a person's behavior is unpredictable even if the behavior of all the other matter in the known universe is predictable. In other words, *a believer in* **OFW** *may maintain that the behavior of humans is beyond the laws of science since it is produced by supernatural causes.* However, most scientists and philosophers who study the mind do not find this approach even slightly appealing.

DETERMINISM, DECISIONS, AND NEURONS

Let's discuss how determinism is in conflict with **OFW**. Since I am discussing this from the point of view of science, I will ignore minor miracles. Consider a decision that you make. A World→Brain→ World interaction is often involved when you make a decision and take an action. That is, a situation in the outside *world* presents itself to you through your senses. In your brain, different courses of action are considered. An action is selected by your brain and it sends nerve signals to muscles that carry out the selected action. This action affects the world. Thus, what starts in the *world* is processed by your *brain* and produces your action on the *world*.

Now let's consider how determinism affects your brain in this picture. Your brain is a physical object and what it does is determined: (1) by the detailed situation at the time; and (2) by the laws of nature that apply to matter in your brain just as they do to all other matter. Consequently, your brain's selection of a certain course of action can happen only one way, given the laws of science and all the circumstances. *Your decision is the only possible one, given the exact circumstances of the situation. Similarly, the action that you take is the only possible action that you could have taken under the precise circumstances.*

We can also regard your decision as a brain process that results from the operation of all the neurons that are involved. Each neuron produces output signals in response to a combination of factors. These factors are the nature or properties of the neuron, its environment, and the input signals that affect it. The neuron's nature determines its response to the other factors. The neuron's environment includes its recent activity, the temperature of the surrounding matter, and the concentrations of chemicals that feed the neuron or affect how the neuron responds to signals. The input signals consist of pulses from other neurons that arrive at specific times and locations on the neuron. Together, these factors control the output of the neuron.

All the factors that affect the output of your neurons are part of the detailed circumstances that exist during the process in which you make your decision. Under the specific circumstances that are present, *each neuron can only do what it actually does*. Your decision results from the operation of all the neurons and it is the only decision that could have been made under the actual circumstances.

So, if we accept determinism and ignore the possibility of minor miracles, a choice that you made was the only choice that you could have made. There was no arbitrariness in this choice. This contradicts the idea of **OFW**, which claims that more than one outcome of your decision process was possible under the actual physical circumstances. Thus, *if determinism is true and there are no minor miracles, then* **OFW** *is false.* If determinism is true, you do not *originate* actions, since your action's original causes were already present in the detailed circumstances that existed before the decision was made.

FREE WILL AND CHAOS

In the previous section I argued that the determinism of classical physics is not consistent with **OFW** unless minor miracles really happen. Modern physics has produced two well-supported theories that challenge the rigid determinism of classical physics. These are *chaos theory* and *quantum theory*. They need to be discussed since we may suspect that they affect the free-will issue. Chaos theory will be discussed first.

Let's define chaos theory. Just for laughs, a technical definition is that *"Chaos theory is the qualitative study of unstable, aperiodic behavior in deterministic nonlinear dynamical systems."*[10] This concise definition contains a great deal of information, but needs to be translated into ordinary English. What it means, for our purposes, is that chaos theory describes the behavior of some *deterministic* physical systems that have erratic behavior. The behavior of such systems is effectively impossible to predict far in advance. Two examples are Earth's weather and the highly fluctuating populations of certain insects. Chaos theory tries to find simple ways of describing the seemingly disorderly behavior of such systems.

The systems that chaos theory studies exhibit what is called *chaotic behavior*. These systems can undergo major changes that are produced by extremely tiny causes. An example is the so-called *butterfly effect*. It suggests that the flap of a butterfly's wings in

Brazil may change the course of the weather so that, later, a tornado might be produced in Texas.[11] Quite a bit has been learned about chaotic systems, *but what is of immediate interest to us is the unpredictability of these systems and the fact that they are deterministic.*

Since chaotic systems are deterministic, scientists might have expected that their future behavior could be predicted using scientific laws and the system's initial conditions. Scientists are used to situations where small, unavoidable errors in measuring initial conditions do not have a large effect on the future of the system. For chaotic systems, however, a tiny error in measuring the initial conditions causes the *predicted* and *actual* behavior to differ by an amount that grows extremely rapidly with time.

The extreme sensitivity of chaotic systems to the *exact* values of the initial conditions makes it impossible, in practice, to make useful long-range predictions of the future behavior of the system. We can imagine that this might apply to the human brain. If so, it may never be possible to make accurate predictions of what a person is going to do. Now consider how this affects the free-will issue.

A person who must make a choice approaches that choice with some specific set of *initial conditions.* That is, at some time before the decision is made, there is a precise answer, in principle, to every question that could be asked about the state of the person's entire brain, body, and surroundings. Even if the brain is a chaotic system, determinism implies that *the decision that is made is the only decision that was possible,* given the actual initial conditions. This contradicts **OFW**, which asserts that a person *might have made another choice* under exactly the same circumstances.

The fact that even a chaotic system is *deterministic* implies that there is only one possible outcome of a decision process. The fact that the resultant decision was not, in practice, *predictable far in advance* does not affect this conclusion. The conclusion is not affected by the fact that no one could have had perfect knowledge of what the initial conditions actually were. The existence of chaotic systems does not make **OFW** any more believable than it was from the perspective of classical physics.

FREE WILL AND THE QUANTUM LOOPHOLE

The way most physicists interpret it, *quantum theory* implies that determinism is not true. Determinism was the source of the conflict between **OFW** and classical physics, so one may ask whether quantum theory allows **OFW** to be in harmony with modern physics. In other words, is there a *quantum loophole* that allows **OFW** to be consistent with what is now known about nature? Many people who are convinced that *some kind* of free will exists have felt that quantum theory may provide such a loophole. Werner Heisenberg mentioned this possibility in 1927.[12]

For our purposes, only a few consequences of quantum theory need to be discussed here. You may read Appendices I and II if you want to learn a little more about quantum theory.

Classical physics doesn't work for extremely small physical systems, such as atoms. Quantum theory, however, has been successful in describing such tiny systems. The microscopic world seems to follow some strange rules that are part of quantum theory. One of quantum theory's strangest results is that one usually cannot predict exactly what will happen in events involving extremely small objects. Nature introduces some randomness into the world. For example, we can't predict when a particular radioactive nucleus will break apart and emit some radiation. All we know is that the nucleus will almost never survive for more than, say, ten times the average lifetime for that kind of nucleus.

Still, there is a method to this quantum madness. Although a particular nucleus will break apart at some random time, the decays of a huge number of nuclei follow a precise pattern. Where large numbers of nuclei are involved, *statistical laws* govern how many nuclei break apart within a certain time interval. Since the nuclei obey these statistical laws, the process is not completely *arbitrary*, even though it is *random within certain limits*. In all kinds of situations where quantum theory is used, things happen this way. Nature usually introduces some randomness, but it always follows certain overall patterns.

Most physicists believe that the randomness in microscopic physical processes arises in nature itself. It is not a result of human ignorance of initial conditions or side effects of experiments. This is not like the randomness in flipping a coin or throwing dice. In principle, by knowing the necessary physics and precisely controlling the initial conditions, one could predict the outcome for a flipped coin or tossed dice. These nonmicroscopic objects are large enough to operate according to nearly perfect determinism. Their behavior should be predictable.

If we combine chaos theory with quantum theory, we can imagine that some large-scale systems may be so sensitive to tiny causes that a random quantum effect, such as the decay of an atomic nucleus, could change the future behavior of the system. By this means, a nucleus's random breakup might cause a neuron in a butterfly to pulse, causing the butterfly to flap its wings, eventually leading to a tornado in Texas. Such situations may cause even large-scale processes, such as Earth's weather, to be *unpredictable in principle, no matter how well we measure the inital conditions*. Even with the help of superhuman intelligence, *it would be impossible to predict the behavior of such processes over long periods of time*.

In a person, a quantum effect might affect a neuron, thereby changing a person's decision. *If we could control these effects*, we might imagine that they could make **OFW** possible, as in Eccles's model that was discussed in chapter 11. There are at least three major problems with this idea. One problem is that it appears that *we have no control over random quantum processes*. Science has found no "handle" by which random quantum processes can be manipulated. There probably *is* no such handle, and it is extremely unlikely that mental processes could move such a handle.

Another problem is that, if we believe that the brain's activities produce mental processes, it makes no sense for one's mind to wish to control the course of a brain process by manipulating quantum effects. If one's mind prefers some course of action, then *brain processes have already selected this course of action*. From this point of view, *the brain and the mind do not have different preferences*: they are connected as closely as a dog and its bark. It seems that *believing in* **OFW** *is linked to a worldview in which the mind is independent of the brain, so that the mind follows its own course and then tries to impose its preferences on the brain.*

The third problem with linking **OFW** to quantum effects is that *random quantum processes obey precise statistical laws.* Presumably, decisions controlled by **OFW** do not obey statistical laws. As a result, any *quantum processes controlled by free will would not obey statistical laws.* The arbitrariness of the behavior that would result from **OFW** does not match the random, but still orderly, properties of quantum processes. This leads me to expect that *the outcomes of random quantum processes in the brain are not arbitrary and not controlled by a supernatural human will.*

In the Eccles model discussed previously, quantum processes have *a very small effect* on the behavior of neurotransmitter packets. The insignificance of quantum effects on the neurotransmitter packets has been verified by Euan Squires, a British professor and author who is studying quantum theory and consciousness. He found that quantum effects would alter the release of only about *one of each four thousand* neurotransmitter packets.[13] Eccles overcame this difficulty by proposing that *hundreds of thousands of neural synapses are simultaneously influenced* by the mind. As a result, enough packets are affected that the outcome of brain processes could be altered.

If we ignore minor miracles, *the quantum processes in the hundreds of thousands of synapses that are involved in a human decision would act at random.* Most synapses would not be affected by quantum effects. Some would be affected in favor of a certain decision outcome, while others would be affected in favor of another outcome. *On the whole there would be no important effect of quantum processes on the behavior of the human nervous system.* In any case, there is no reason to believe that the "mind" can control quantum effects. Thus, quantum effects are hardly any more compatible with **OFW** than determinism.

If we consider that the nervous system is a product of biological evolution, we would expect that it would function reliably. It should consistently allow humans and animals to make decisions that lead to the greatest probability of their survival. *Such decisions should not be made at random.* As an example, think of animals that live high in trees and jump from branch to branch without safety nets. So, nervous systems should not be very sensitive to random quantum effects. This is another reason for believing that quantum effects have little influence on decisions of humans or animals.

In the previous two paragraphs, we assumed that minor miracles are not common. If there *are* minor miracles, they are potentially detectable, at least in principle. Any case in which the outcome of events in the real world is actually changed due to **OFW** is detectable in principle. *If* **OFW** *is true, it could eventually be detected by science.* For now, there is no scientific evidence favoring **OFW**. In the next chapter, we shall see that one's subjective feeling of making one's own decisions is not a strong reason to believe in **OFW**. This feeling will be explained in another way.

In discussing quantum processes, I have followed the opinion of the majority of physicists. Other ways to interpret quantum theory have been suggested.[14] One interpretation, usually regarded as fairly unlikely, is that quantum processes are not intrinsically random. The underlying processes are deterministic in this view, involving what are called "hidden variables." I have already concluded that determinism is contrary to **OFW**, so this approach does not change our main conclusion.

Some readers may know of another interpretation of quantum theory that is called the "many worlds" picture. Even in this view, the particular branch of the universe that we inhabit behaves with the randomness that I have assumed above. This picture supports the same conclusion on **OFW** that has been reached here.

CONCLUSIONS ON OBJECTIVE FREE WILL

Let's summarize the discussion of chaos, the quantum, and their effects on **OFW**. The presence of random quantum effects in chaotic systems implies that determinism is not true. *Some* of what happens in the world is not predictable far in advance, even in principle.

Quantum effects probably do not play a large role in affecting human decisions. Also, quantum effects appear to be intrinsically random and *not controllable by the mind*. If we assume that quantum effects really are not controlled by the mind, human mental processes are still completely determined by natural causes, even if some of these causes are random quantum effects. *If* **OFW** *is true, it seems to require the existence of minor miracles that are outside the known possibilities of any broadly accepted physical theory, including quantum theory. There seems to be no convincing reason to think that quantum theory supports the possibility of* **OFW**.

Objective Free Will is inconsistent with determinism and is not supported by chaos theory or quantum theory. It has been noted that "All known processes of the physical world are either deterministic or exihibit quantum randomness."[15] Since neither determinism nor intrinsic randomness supports **OFW**, it appears that **OFW** does not exist unless minor miracles are very common. If minor miracles happen, science should eventually detect them. So far, they have not been detected.

OFW is an *anthropocentric idea*. That is, it assumes that we occupy a unique, elevated status in the universe. **OFW** supports the view that our mental processes violate the orderliness that is present throughout the rest of the universe. According to **OFW**, the matter within our brains operates differently than all the other matter in the universe. **OFW** may eventually join other similar ideas that have been overthrown or made doubtful by scientific progress. These include such ideas as that: Earth is the center of the universe; Earth does not move; God's intervention is necessary to keep the planets in their proper orbits; life did not arise from purely natural processes; and we are not descended from any other animals.

I have argued above that free will is not a supernatural process. Free will should probably be excluded from the alleged powers of the supernatural soul. Although nothing remains of the supernatural soul, I am inclined to believe that there is nothing to mourn. Denying the supernatural soul's existence does not renounce any real human properties *if nobody ever had a supernatural soul.* All of our psychological properties can probably be explained by the *natural soul* of chapter 11.

NOTES

1. Albrecht Dihle, *The Theory of Will in Classical Antiquity* (Berkeley: University of California Press, 1982), p. 105.

2. Ibid., p. 41.

3. *The Encyclopedia of Religion* (1987), "Free Will and Predestination," by W. Montgomery Watt.

4. Dihle, *The Theory of Will*, p. 123.

5. Augustine, *On Free Choice of the Will* (Indianapolis: Bobbs-Merrill Co., 1964), p. 3.

6. Dihle, *The Theory of Will*, p. 133.

7. Albert A. Michelson, quoted in John Bartlett, *Bartlett's Familiar Quotations*, 14th ed. (Boston: Little, Brown & Co., 1968), p. 827.

8. A. C. Crombie, *Medieval and Early Modern Science*, 2 vols. (Garden City, N.Y.: Doubleday, Anchor Books, 1959), 2:47.

9. Roy Weatherford, *The Implications of Determinism* (London: Routledge, 1991), p. 56; Bertrand Russell, *Religion and Science* (New York: Oxford University Press, 1961), p. 58.

10. Stephen H. Kellert, *In the Wake of Chaos* (Chicago: University of Chicago Press, 1993), p. 2.

11. Ibid., p. 12.

12. Nathan Spielberg and Bryon D. Anderson, *Seven Ideas That Shook the Universe*, 2d ed. (New York: John Wiley & Sons, 1995), p. 276.

13. Euan Squires, *Conscious Mind in the Physical World* (Bristol: Adam Hilger, 1990), pp. 221–23.

14. Victor J. Stenger, *The Unconscious Quantum: Metaphysics in Modern Physics and Cosmology* (Amherst, N.Y.: Prometheus Books, 1995), chaps. 3–7.

15. Jean E. Burns, "The Possibility of Empirical Tests of Hypotheses about Consciousness," in *Toward a Science of Consciousness: The First Tucson Discussions and Debates*, ed. Stuart R. Hameroff, Alfred W. Kaszniak, and Alwyn C. Scott (Cambridge: MIT Press, 1996), p. 741.

14

FEELING FREE AND BEING FREE

Man's life is a line that nature commands him to describe upon the surface of the earth, without his ever being able to swerve from it, even for an instant. He is born without his own consent; his organization does in nowise depend upon himself; his ideas come to him involuntarily; his habits are in the power of those who cause him to contract them; he is unceasingly modified by causes, whether visible or concealed, over which he has no control, which necessarily regulate his mode of existence, give the hue to his way of thinking, and determine his manner of acting. He is good or bad, happy or miserable, wise or foolish, reasonable or irrational, without his will being for anything in these various states.

—Paul-Henri Thiry, Baron d'Holbach[1]

SHAPED BY THE UNIVERSE

BEFORE DISCUSSING IN WHAT WAYS we are free, I would like to describe the way in which all of our behavior is determined by causes that come from outside ourselves. In this section, I will take the word "responsible" to mean that one is not "responsible" for things that are ultimately and completely determined by causes outside of oneself. Using this meaning of "responsible" makes sense if one is thinking of a certain rigid kind of moral responsibility, for example. We shall see other ways in which the word "responsible" can be used in the following chapters. I use quotes around the word "responsible" throughout this section in order to remind the reader that I am using the word in this particular and restricted sense.

Let's consider, in a very general way, how one's brain develops. This is relevant because the brain, affected somewhat by the other parts of a person's body, determines the behavioral characteristics of a person. A person's knowledge, memories, motivations, values, character, and mental abilities are mostly determined by the enormously detailed structure of the brain.

Initially, a fetus has no brain. Without a brain, the mental properties of knowledge, memories, consciousness, and so on are absent. When the brain develops, its formation is guided by the information contained in the developing human's genetic code. Obviously, the developing person is not "responsible" for the information contained in that code, nor for the environment of the womb, which also affects the development of the nervous system. It is up to the mother to avoid exposing the fetus to harmful drugs and the effects of alcohol. Until birth, the baby has no control over what happens to it.

At a very early stage, perhaps even within the womb, the learning process begins. As a result of the person's experiences, connections are made and strengthened in the brain. These connections are made according to orderly physical processes. Just *which* connections are made and enhanced depends on the details of the experiences of the young person. What one learns is somehow encoded by these connections. The content of what one learns is determined by one's experiences. At this very early stage, one is not "responsible" for one's experiences, so one is not "responsible" for what is learned.

One's earliest *beliefs* are determined by what one has learned.

At this stage one is not "responsible" for what has been learned. Consequently, one is not "responsible" for one's earliest beliefs. One's earliest *values* are determined by one's early beliefs and experiences, together with one's *innate values* that are determined genetically. An example of such an innate value is the feeling that pain is bad and is to be avoided. One is not "responsible" for one's earliest beliefs and values.

A little later, new beliefs are based on one's earliest beliefs and one's new experiences. Similarly, one's new values are based on one's earliest values and beliefs, combined with one's new experiences. One is not "responsible" for these earliest beliefs and values or the content of these new experiences, so one is also not "responsible" for these new beliefs and values. At each stage in the learning of new beliefs and values, one is not "responsible" for one's previous beliefs and values. One is also not "responsible" for one's new experiences. Consequently, one is not "responsible" for one's new beliefs and values. *Thus, one is not "responsible" for one's current beliefs and values.*

At this point someone might object and say that one can have some *control over one's new experiences*. One may, for example, choose to read certain inspiring religious literature, so that one will benefit from learning the values contained in that literature. Obviously, such behavior can be used to control one's experiences and what one learns.

Note, however, that one is usually likely to read inspiring literature only if one already holds beliefs and values that are consistent with what one expects to find in that literature. Thus, one's choice to read the literature is determined by one's previous beliefs and values. Since one was not "responsible" for those beliefs and values, one is ultimately not "responsible" for exposing oneself to the presumably beneficial effects of reading the inspiring literature. As a consequence of this, one is also not "responsible" for what one learns from the inspiring literature. All of this can be traced to causes outside oneself.

One's beliefs and values, together with factors related to one's genetic makeup, result in one's *character*. One is not "responsible" for one's own character or motivations since one was not "responsible" for the causes that shaped one's character. A person's character, beliefs, and motivations, along with external circumstances, determine one's behavior. One is ultimately not "responsible" for the determining factors of one's behavior. Therefore, there is an important sense in which *one is ultimately not "responsible" for one's own behavior*. One's behavior is ultimately determined by the effects of outside causes. In other words, one's behavior is shaped by the universe.

Our discussion has led to a startling conclusion. It appears that there is a real sense in which *one is not "responsible" for one's beliefs, values, character, and behavior*. This conclusion may seem to be unacceptable to many readers. I should add, however, that it is still possible to *hold a person "responsible"* for his or her actions, and a person may hold himself or herself "responsible" for his or her actions. This will be discussed in the following chapters. For now, let's consider another argument leading to similar conclusions about our ultimate responsibility for our own actions.

Consider the string of decisions a person makes during her lifetime. These decisions determine her behavior. I concluded earlier that when each decision is made there is only one outcome that can be selected under the exact circumstances that are present. The person makes the only decision that is possible for her. As a result, *the person's decision leads to the only behavior that is possible for her*. It does not seem fair to hold the person strictly "responsible" for her behavior, since she was really not able to do anything else. She did not *originate* her decisions and actions, they were the natural result of physical processes that originated outside of herself.

It might be argued that she could have behaved better if her character were better, and that she is "responsible" for her character. In reply, I can ask how she could have chosen a better character. Each choice she ever made was the only choice that was possible for her at that time, given the circumstances. So, someone is being somewhat arbitrary, harsh, and ignorant to hold her "responsible" for what her character has become. Of course, by the same argument *we* shouldn't be too critical of someone who is arbitrary, harsh, or ignorant.

The argument may be more convincing if it is made in terms of the neurons that make up the nervous system of the person. Each decision in this person's life comes about as a result of the operation of her neurons. Each neuron operates automatically, according to the physical laws that regulate its functioning. The behavior of each neuron is determined by the physical laws and the circumstances at the time of the decision. Each neuron functions the only way it can, so that the person's entire nervous system functions in the only way it can. As a result, only one decision is possible for the person. The person can behave in only one way. Consequently, there is a real sense in which it is unfair to blame her for her behavior.

If I held a more traditionally religious view of human nature, I might say that the person's character is determined by the quality of her soul, but I have argued against the existence of a soul or minor miracles that would allow her to make arbitrary choices in her moral decisions. Instead, it seems clear that the person's character is determined by the detailed structure of her nervous system. This structure was formed entirely by her genetics, the relevant physical laws, and the circumstances that existed throughout her earlier life.

Assuming that the physical laws are always applicable, and that the person can do nothing about her genetics, the only remaining influences on the person's nervous system come from the environment. The environment affected the structure by providing suitable growing conditions for neurons and by guiding the connections between neurons in the learning process. If we are trying to be "fair" to a person, the person should not be blamed or praised for any part of the process by which the structure of the nervous system developed. The person was really not "responsible" for how this process developed. Again, I conclude that the person is not "responsible" for her character or behavior, in that her character and behavior were ultimately and completely shaped by outside causes.

For a number of reasons that I will discuss in this and the following chapters, this conclusion is not as unfortunate as it seems. We still maintain *control* over our behavior. We still can have values and can live according to them. We can still be *held responsible* for our behavior. *Nevertheless, it is somewhat unfair to discriminate between people on the basis of their behavior, for which they are ultimately not "responsible."* This sort of fairness, however, is not our only value. To some extent *we are willing to sacrifice some of this kind of fairness in order to secure other goals, such as survival, prosperity, and an orderly society.*

FREEDOM AND LIMITS TO OUR SELF-KNOWLEDGE

We all feel free to make choices. When we make a choice, it seems that we choose between different options that are really possible. How can we reconcile this with the discussion of the previous chapter, where it was concluded that one makes *the only choice that is possible under the exact circumstances that apply?* Also, how can we reconcile this with the previous section, in which I concluded that one is not ultimately "responsible" for one's own behavior. If that is true, then how meaningful can our choices be?

To answer the questions asked above, we need to distinguish between two things. One is the physical world, in which only one outcome of a decision is normally possible. The other is one's subjective experience, or consciousness, which is an incomplete representation of the world. A person has incomplete knowledge of what is going on in the outside world. Of more importance to this discussion, one has incomplete knowledge of *what is happening within one's own brain.* One does not know which choice will be made and one does not have precise knowledge of *why* one made a particular choice.

Since one does not know which choice will be made, the different options appear to be possible choices, even though only one choice is consistent with the detailed physical situation and any random quantum effects. One does not know what the precise details are, or what random quantum effects may occur. Even if one knew all these details, the process of weighing all these influences and making a choice would still need to be done by the brain. The brain comes to the only decision it can make, and then we become aware of the decision. Sometimes we are surprised by which decision was made, or that the decision came at the time that it did.

One never knows all the details that affected the decision. First, let's consider this at the level of the neurons. Perhaps the smallest items that may enter our awareness are individual pulses from neurons. We are not aware of such details as how input pulses at various locations on a neuron will combine to affect the probability of the neuron's firing. We may not be aware of changes that may occur in the trigger level at which a neuron will fire a pulse. *But these details may be of vital importance in determining the outcome of the brain's act of choosing.* The timing and the outcome of a decision are often not predictable by a person, since many of these essential details are not available to our consciousness.

Now let's shift levels from the individual neurons to the overall mental processes that may or may not enter our consciousness. Before making a careful choice, we are aware of reasons in favor of or against each option. The process of choosing may take into account all the implications of the different possible choices. Thus, we make our choice subject to all of our stored knowledge concerning the likely results of our choice. This weighing of the implications results in a *meaningful* choice.

One's conscious mind is not competent in predicting the actual decision outcome, since many important details may not be available to one's consciousness. Also, some of the influences on the decision may be *subconscious motivations*. These, of course, are not available to one's awareness. Making a decision is a sort of mechanical process in which various relevant conscious *and subconscious* motivations affect the outcome. As a conscious person, one submits the problem, with all its implications, to a decision-making mechanism. One can't predict the outcome, but one is confident that all of one's motivations and considerations are taken into account in the decision process.

When a decision "arrives" in our awareness, we often need to explain our decision to ourselves or other people. For complex psychological reasons, one's mind works to put a favorable "spin" on the decision, *even to oneself*. Reprehensible reasons for choosing an action are sometimes left out of the explanation of the choice.

The mind can explain the decision only in terms of the reasons of which it is aware. It constructs an explanation of the decision that is consistent with known reasons that were in favor of the actual choice. Unless they can be inferred, unknown details and subconscious motivations that affected the process are left out of the explanation. *So, a person always has a limited grasp of why a specific choice was made.* Even a very truthful explanation of a decision may be misleading. This picture of the effects of our limited knowledge on our subjective feeling of freedom is quite similar to an approach taken by Francis Crick.[2]

Under the circumstances which have been described above, how can we expect decision making to seem to a person? *To a person who is about to make a difficult decision, the different options all seem to be possible choices.* A philosopher or scientist may claim that only one decision will be possible when the decision is made. To the person making the decision, however, the decision is up to her, and more than one outcome seems possible.

When one makes a decision, it seems that the decision was one's own. Although one was not aware of all the details or motivations leading to the decision, it was made according to one's own motivations and reasons. The decision was made in close connection with one's awareness at the time when a decision was needed. One became aware of it shortly after it was made. Normally, no one else is aware of the decision until one informs someone. The decision is consistent with one's desires and it seems that one actively chose the outcome, since the choice followed one's own efforts in making the deliberations.

When a decision is made, one seems to have control over which choice was made. If one divides the universe into oneself and everything else, one must admit that the choice came from oneself. The choosing took place entirely within oneself, even if one was unaware of some of the details of the process. In one's awareness, one normally makes this distinction between oneself and everything else. It seems that one controlled the choice, since the decision took place within oneself. If one had *"wanted to,"* one would have made a different choice. *One feels in control of the situation, even if one lives in a deterministic world in which everything that happens is as inevitable as the falling of a dropped stone.*

Another reason why it seems that we control which choice is made is that the choice that is selected usually seems reasonable to us, at least at the time it is chosen. This is natural since we are aware of our reasons and we feel our motivations for making the choice that is made. Our subconscious motivations probably contribute to our *intuitive feelings* about which choice is best for us. In any case, the decision is usually consistent with our feelings or our reasoning about the merits of the different options. Perhaps in most cases, *both* our reasoning and our feelings support the decision that is made.

Another reason why we feel like we control our choices is that this is the usual belief in our culture. Not only are we believed to be in control of our actions, but we are also held to be responsible for our actions. *This is built into our laws, our morality, and our language.* If someone claims that she is not responsible for her actions because everything about herself has been shaped by outside causes, she is usually not taken seriously.

IS FREE WILL AN ILLUSION?

In the previous chapter I argued that Objective Free Will (**OFW**) is possible only if minor miracles are an everyday occurrence, and there is no convincing evidence that minor miracles exist. I also argued that **OFW** is not consistent with classical *or* modern physics. On those grounds, I contend that we do not possess **OFW**. In the previous section, however, I showed that *it seems to us that we do have some form of free will.* Part of the reason that we believe that we have free will is our lack of knowledge of many details of a situation. When one makes a decision, these details eventually cause a particular option to be chosen. If we had perfect and complete knowledge of the entire situation, we would *not* believe that more than one choice is possible. So, part of the usual belief in free will is illusory.

One approach to free will is to focus on the part of the free-will belief that is probably mistaken. One may focus on the belief that more than one option is actually possible when we make a decision. By considering just this aspect of the free-will issue, one may conclude that free will is merely an illusion and that it is *incompatible* with determinism.

From the discussion in the last section, it is apparent that we have a strong feeling that one kind of free will exists. That is the feeling and belief *that we do make choices, that we can make our own decisions, and that we control the outcome of the decisions. This form of free will is part of our personal awareness.* By believing in this form of free will *we tend to feel that we are free,* even if our abilities, motives, and reasons can all be traced, ultimately, to causes outside ourselves. By having this belief, one feels that one's mind imposes its control over matter. It is so commonly present in people that it deserves a name. I will call it *Subjective Free Will* (SFW).

I do not mean to diminish the importance of SFW by labeling it *subjective.* In much of this book we are discussing things that are subjective, such as consciousness. These things have their immediate effect on only one person. They affect the rest of the world largely through their effect on the actions of that person. Although they are subjective, they are real. Few would deny that consciousness is real. We are aware of the objective world only through our subjective consciousness. Thus, each scientist depends on something that is subjective for knowledge of the "real, objective, world." In addition, if science is going to deal satisfactorily with mental processes, it will probably need to include subjective phenomena in its worldview.[3]

Are the set of beliefs that make up SFW true? It appears that some parts of the beliefs are true and that other parts are false. Let's discuss what seems likely to be true or false about SFW. Consider the feeling that we have choices. When we make a decision and choose a particular option, the chosen option is the only choice that is possible at the time the decision is made, given all the deterministic causes and random quantum effects that made our decision-making mechanism act the way it acts. *If having a choice means that more than one option could have been selected with things exactly as they were when the decision was made, then we do not have choices.*

Before a decision is made, however, we do not know what option will be selected by the decision-making mechanism. We must consider the options and go through the process of making a choice. In the sense that the decision *depends on what one chooses*, one has a real choice. As a practical matter, it is reasonable to believe that one has a real choice, since one must prepare to make a decision. *If making a choice means considering options and going through a decision process of which the outcome is not known in advance, then we do make choices.*

Now let's consider whether "our" choices are really "our own." When we walk down a street, we take with us all that is needed to make our choices. In this sense, the choices are made within ourselves. Our choices are made according to our values, our motivations, and our own character. Our choices determine our own actions. Society holds us responsible for our choices. *A choice can be called one's own if it is made by one's brain's normal decision processes and without extreme outside interference. In this sense, we make our own choices.*

From another point of view, however, our choices result from predetermined and random causes. The universe always operates according to these causes. They are universal processes, and are in no way "our own." Also, we do not choose our genes or our early experiences. We are what we are because of a continual chain of causes that come from outside ourselves. Our values, motivations, and character can all be traced to outside sources. *Since our choices are selected due to things that ultimately arose outside ourselves, it can be argued that the choices that we make are not really our own.*

Do we have control over our decisions? As above, we may note that everything leading to the decision is shaped by the universe. However, the universe does not *control* our decisions, in the sense that *control* involves some intention to produce a certain outcome. The universe functions the only way it can, but does not *intend* to do anything.

We can predict how things will go if we choose various options.

With these predictions, we can select options that appear likely to enable us to reach some goal. In such a case, the decision outcome is largely determined by our goals and our analysis of the situation. *An intelligent being has the ability to control decisions* in this way. This ability to obtain rational control over the environment is an emergent property of an intelligent organism. Daniel Dennett has a very helpful discussion of control and self-control in his excellent book, *Elbow Room*, which is devoted to the free-will issue.[4]

The belief in free will plays a large part in our lives. Some commonly accepted parts of this belief are illusory, while other parts of the free-will idea are true. As intelligent beings, we can make decisions that give us effective control over some of our surroundings. Our values and wishes can have a real effect, even in a world that features large-scale determinism and small-scale randomness. *It would be a great mistake to deny the parts of free will that are true. Through exercising our ability to control events, we affect the world in the most meaningful way that is possible, given the nature of the world.*

THE EVOLUTION OF FREEDOM

I use the word "control" *in a looser sense* here than I used it in the previous section. Here, the word is not always limited to actions by humans. In this loose sense, the temperature of a house can be controlled by a thermostat. Electronic and mechanical devices make decisions, as when the thermostat "decides" to turn on the furnace. These are examples of choice and control exercised by a mechanism that lacks awareness.

In this same loose sense, choice and control are intimately linked with the evolution of animals. Control is a very basic feature of animal life. By using sensory organs, nerves, and muscles, animals are able to select and carry out various movements. By these means animals are able to control their location and many other things, and because of this they became a very important part of life on Earth. Some roots of our freedom to make choices go back to our animal ancestors that lived hundreds of millions of years ago.

Subjective Free Will, however, was not present in the first animals to have control of their motions, because SFW requires consciousness. The consciousness that is required for SFW goes beyond awareness of things that affect our senses, since it involves a belief about the nature of a mental process: decision making. SFW requires that its possessor be aware of at least part of the process of decision making. Awareness of one's mental processes is considered to be a higher form of consciousness than sensory awareness. On the other hand, it is not hard to imagine that a brain that can represent what happens in the outside world could also represent some of its own internal processes. This could be accomplished without any new sensory organs, "just" by making additional connections within the brain.

One can imagine how awareness of the mental activities *of others* could arise in intelligent social animals. Some primates, for example, may have improved their ability to live successfully in groups by learning to anticipate the actions of others in their group. This may have led them to be aware of others' feelings and motivations. The decisions of others might have been predicted by imagining what the other animal knew or felt. By learning to understand the behavior of others, a crude idea of the minds of other animals might have developed. *It may have been more useful to understand the minds of others than to understand one's own mind.* One's own behavior could be guided by feelings and intuition, while one needed to *infer* what motivations others had, so that one could predict their actions.

After animals developed ideas about the minds of others, they may have taken another difficult step and realized that they, themselves, have minds. They may have learned to identify certain feelings as emotions or motivations that lead to certain kinds of behavior. By this route, our ancestors may have developed awareness of their own mental processes. On the other hand, the awareness of one's own mind may have developed first. It is valuable to be able to predict one's own future behavior. It is also valuable to understand why one believes something, so that one can know how much confidence to place in this belief.

One of the early things that our ancestors learned about themselves and others may have been that individuals have choices and make decisions. These are basic since they lead to actions that are essential for survival. That humans can control their environment was learned very early, leading to the use of structures and fire. A belief in one's ability to make decisions and to control the environment may have led to a belief in some kind of free will even in prehistoric times.

Our ability to make choices has very old roots and arises from our nature as animals. Because of this, it seems somewhat perverse to claim that free will is a supernatural rather than a natural ability. In this and some other cases involving our fundamental nature, *we have a strong tendency to claim a status that is higher than our nature*. For example, we would rather believe that we are angels burdened by bodies than that we are descended from apes.

ACCEPTING LIMITED FREEDOM

Some of what we believe about ourselves and other people will change if we accept the ideas about free will that were described above. Before discussing the *changes*, however, let's consider what stays the same if we embrace these unorthodox views about free will.

The common belief in Objective Free Will (**OFW**) implies that minor miracles occur frequently, allowing us to do things that would be impossible if nature followed its normal course. The belief in minor miracles does not make it possible for them to occur. If there really *are* no minor miracles, we can't bring them about by believing in them. We can ignore them without neglecting any possibilities in the real world. If one concludes that the belief in minor miracles is misguided, one can take a more realistic view of the world by discarding one's previous belief in Objective Free Will.

Although it is reasonable to deny **OFW**, it is not wise to reject some other things that are associated with our ideas of free will. Although we do not exercise supernatural powers, we continue to possess such valuable mental assets as knowledge, intuition, and intelligence. We have the ability to plan our actions and make prudent decisions. We can carefully control our own actions, so that they become as consistent with our values as is possible.

Although our choices are determined by past and present circumstances, there is no reason to embrace *fatalism* regarding our future actions.[5] Fatalism is the idea that everything that happens to us will happen because it is our destiny: *what we do makes no difference*. This idea tends to defeat one's ability to reach certain goals by working to accomplish them. Clearly, we will not reach our greatest potential if we do not work to achieve our goals. In many situations, our actions can make a crucial difference in what happens. We can maintain an active stance toward the world and try to steer things in the right direction. We are not helpless pawns of the universe. We are parts of the universe equipped with intelligence and muscles. If the world is to be improved, we are the most plausible means of accomplishing this.

Part of the free-will idea is true. That part includes our ability to control our actions and make a difference in the world. It also involves our ability to plan our actions and make careful decisions. *The fact that our decisions are determined by circumstances does not detract from the feeling of freedom that we have in choosing our actions.* Even when we know our decisions are determined by circumstances, we can remain active in carrying out our role in the world, for we also can be confident that our actions will be guided by our own desires and values.

NOTES

1. Eighteenth-century French philosopher, quoted in Roy Weatherford, *The Implications of Determinism* (London: Routledge, 1991), pp. 93, 94.

2. Francis Crick, *The Astonishing Hypothesis: The Scientific Search for the Soul* (New York: Simon & Schuster, 1994), p. 266.

3. Willis W. Harman, "Toward a Science of Consciousness: Addressing Two Central Questions," in *Toward a Science of Consciousness: The First Tucson Discussions and Debates*, ed. Stuart R. Hameroff, Alfred

W. Kaszniak, and Alwyn C. Scott (Cambridge: MIT Press, 1996), pp. 743–50.

4. Daniel C. Dennett, *Elbow Room* (Cambridge: MIT Press, 1984), chap. 3.

5. Ted Honderich, *The Consequences of Determinism* (Oxford: Oxford University Press, 1988), pp. 16, 17.

Part III

LIVING WITHOUT SOUL BELIEFS

15

TAKING STOCK OF
ONE'S OWN NATURE

Spirit of nature! all-sufficing power,
Necessity! thou mother of the world!
Unlike the God of human error, thou
Requir'st no prayers or praises; the caprice
Of man's weak will belongs no more to thee
Than do the changeful passions of his breast
To thy unvarying harmony; the slave
Whose horrible lusts spread misery o'er the world,
And the good man, who lifts, with virtuous pride,
His being, in the sight of happiness,
That springs from his own works; the poison-tree,
Beneath whose shade all life is withered up,
And the fair oak, whose leafy dome affords
A temple where the vows of happy love
Are registered, are equal in thy sight:
No love, no hate thou cherishest; revenge
And favouritism, and worst desire of fame
Thou know'st not: all that the wide world contains
Are but thy passive instruments, and thou
Regard'st them all with an impartial eye,
Whose joy or pain thy nature cannot feel,

Because thou hast not human sense,
Because thou art not human mind.

—Percy Bysshe Shelley, from *Queen Mab*

THE QUOTATION FROM SHELLEY'S POEM describes a world like the one that has unfolded in the previous chapters. The world in which we live seems to be completely unconcerned with human morality or justice, as if the basic foundations of our existence are impersonal in nature. Perhaps the greater wisdom is to accept the world as it is, rather than to believe in a hidden reality that is much better than the world we know.

Even in such a world, however, it is usually not desirable to live the life of a radical poet. Much of life remains the same even if we accept a worldview that maintains that we are truly mortal and possess only the natural varieties of free will and consciousness. It is time to summarize the gains and losses that we experience if we accept the worldview implied by a natural soul that is endowed with a subjective, but not an objective, free will.

ACCEPTING DEATH AS REAL

Our hopes often arise from our wishes. We usually try to restrict our hopes to wishes that can possibly be fulfilled. Some wishes, however, arise from such a basic desire that they spawn hopes based mostly on wishful thinking. *There is no more basic desire than to survive. This, combined with the desire that our loved ones also survive, leads to an intense hope for eternal life.* For many successful religions, this hope is a key to human hearts; for the hope for immortality is so strong, and eternal life is so easily promised. In extreme cases, everlasting happiness in heaven can be so tempting that suicide must be sternly condemned!

For some of us, the hope for immortality can be held even if it is based on shaky arguments since the principal cause of the hope is the intensity of the basic desire. Still, there is a broad argument for immortality that is intertwined with another assumption that is commonly held within our culture. This assumption is that *mental processes are not part of the material world*, leading to a belief that these processes are of a transcendant or spiritual nature. A belief in a spiritual, or nonphysical, element in human nature suggests that there may be eternal life, since the spiritual parts of our nature are not likely to be subject to the physical processes of death, dissolution, and decay. *Because of the intensity of the desire, the mere suggestion of the possibility of immortality can give rise to the hope for eternal life.*

In previous chapters I have argued that consciousness, conscience, and free will are natural processes in the human nervous system. Although these abilities have traditionally been regarded as faculties of the supernatural soul, there seems to be no need for a supernatural part of human nature to grant us these powers. They arise naturally from the structure of our nervous system. In the case of *conscience*, one's interactions with society help to form the ideas of what is right and what is wrong. These ideas, along with awareness of events, reasoning, and intuition, help one to judge whether one's actions are proper or not.

It no longer appears to be necessary to posit a supernatural part of our nature. Thus, we have no convincing support for a transcendant part of our own nature that might be capable of surviving the death of our bodies. *We are left with the likely truth that death is what it appears to be: the end of conscious life.* Within the Christian system of beliefs, however, there is another way that we might become immortal. This is by the resurrection of our bodies. In this picture, shared with Zoroastrianism, Islam, and parts of Judaism, our bodies will return to life at the time of the final judgment.

The basis for the Christian belief in the final judgment and the resurrection of the dead is the Bible. In chapter 4, it was argued that such beliefs as the resurrection of the dead were not part of the authentic teachings of Jesus. These beliefs were brought into the scriptures by very early followers of Jesus. They coincided with the apocalyptic beliefs of certain groups within Judaism, who may have borrowed them from earlier Persian doctrines. If, however, the beliefs are not the authentic teachings of Jesus, they must lose much of their believability among Christians.

The religious arguments for human immortality seem greatly weakened. From the starting point that our nature is entirely consistent with natural laws, the idea of human immortality appears to be a bizarre concept. The religious factions that promise immortal life seem to be making extravagant promises with little support. There appears to be no more reason that life should continue after we die than that we should have been alive before we were conceived. In our society, at least, few believe that we were alive before we were conceived.

Even if the belief in immortality is likely to be misguided, *we should be careful to respect the views of those who continue to hold that belief*. Opinions about the reality of eternal life are deeply personal and strongly held. In our current culture, we can't expect general agreement on such issues. *The principle of religious tolerance works both ways, however, and we can also expect believers to respect the views of those who do not believe.*

ACCEPTING DEATH WITHOUT FEAR

Although you may resign yourself to the inevitable reality of death, *you do not need to fear death as some sort of terrifying and uncertain form of future existence*. The time when you are dead will be similar to the time *before you were born*. There was nothing terrifying, *for you*, about the billions of years that occurred before you existed. *You had no experiences at all during those years before you were born.* During those years you had no functioning brain that could give rise to any sort of experiences. Similarly, after you die, *there will be no experiences, including the sense of passing time*.

Eternity is less than an instant for a being who has ceased to exist. Death is nothingness, like sleep without dreams. There are no unpleasant experiences to be feared. There is no unhappiness or pain. There is no need to fear being dead. There is no need to ask that the dead "rest in peace." Being dead *is* resting in peace.

There is a sense in which there is no need to prepare for death. One should, of course, get one's affairs in order so that one's loved ones are taken care of after one's death. With all due respect, I submit that it is a waste of one's most precious assets to spend time and effort, while one is alive, in preparing for eternal life. No need to fill pyramids with food and furniture, or to spend one's time in prayer and pious practices in order to store up treasures in heaven. As an evangelist might declare, in a different context, *"These things are all folly!"* Death is the end of the conscious individual. Luckily, however, all of the readers of this book are alive and have some time left to live!

Like death itself, the *process of dying* is not to be feared. People who have nearly died report experiences that are pleasant enough, such as having a departed loved one or a religious figure welcome them into a new realm. I maintain that these experiences have no more reality than dreams, but fortunately they are not unpleasant dreams!

It is not necessarily easier for a religious believer to die than for a skeptic, agnostic, or atheist. Elisabeth Kubler-Ross, a psychiatrist involved with counseling the dying, reported that, "We have worked with only four genuine true atheists and they have died with amazing peace and acceptance, no different from a religious person."[1] She also wrote that, "The significant variable is not *what* you believe, but *how* truly and genuinely you believe."[2] This idea is also expressed by a former priest, Robert Kavanaugh, who is also a teacher and a psychologist. In his book, *Facing Death*, he maintains, "After careful analysis and reflection, I am convinced that belief in an afterlife brings no special peace to the dying that cannot be found by firm adherence to any other belief."[3]

Sometimes it seems that too much is made of how a person dies. Perhaps this is a remnant in our culture of the warrior's creed that death is the true test of a person's character. Dying is not a "final examination" of the quality of one's personal philosophy. How one dies depends a great deal on the nature of the physical cause of death and on what drugs one has access to at the time of death. Dying is an extremely personal experience. Everyone's behavior in facing death should be respected, not judged. If someone prefers to go out screaming and thrashing, we should not be scandalized by that person's behavior.

If someone spent half of his or her life preparing to die peacefully, was all this preparation worthwhile, *or was half of the person's life nearly wasted?* The actual experience of dying occupies only a tiny part of one's life. It does not seem wise to spend much time worrying about it or preparing for it.

Excessive pain and suffering *is* something to fear about dying. Modern medicine, of course, can greatly reduce the pain and suffering associated with death. Another burden of death is the feeling of unfulfilled hopes and the knowledge that these hopes will never be fulfilled. Of course, we have these feelings many other times throughout our lives, and normally we are able to bear them. The feelings of regret about not achieving certain goals are part of life, they are not uniquely a part of death.

Perhaps the greatest burden of death is the loss of the continued companionship of one's loved ones. This afflicts those who are dying and those who are losing a loved one and is one of life's most difficult experiences. Arguments can be made against the *fear* of death, but one can't deny the great *sadness and feeling of loss* that usually accompanies death. For those who survive, it is some comfort to know that the sadness will eventually diminish, but there is no denying that the feelings are intense.

CONTROL AND HUMAN DIGNITY

After discussing death and dying, let's consider how the ideas discussed in this book affect our *lives*. Some readers may feel that believing that all human actions are determined by nature is a challenge to their personal dignity and human dignity in general. It may seem that a person is just "coming along for the ride" in *a universe that is as predetermined as the movements of a merry-go-round.*

In our experiences, however, *it does not seem that we are passive and are just along for the ride.* For example, suppose that you have some difficult work to do, such as cleaning a house or painting a house. If you are tired, it may seem necessary to "push yourself" to do the work. There is not a feeling that the work just gets done because it was predetermined that you would do that work on that day. Although one's desire to finish the work may overcome the tired feeling in one's muscles, it may be a continual struggle to do the work. One may finish the work by "sheer will-power." One may need to deliberate again and again about whether to finish the work or to stop working. It may be necessary to keep one's attention on the reason for finishing the work, so that one's intellectual motives can control one's biological desire to rest.

In the example given above, different parts of the body are in conflict with each other. Tired muscles send one message to the brain, while one's goal to finish the work carries the opposite message. The intellectual goal may prevail while producing minor damage to the body, resulting in soreness and aching joints later. A struggle is occurring that is taking much mental energy. All of this can happen by ordinary physical processes that do not involve minor miracles. My point here, however, is that one's consciousness perceives oneself as taking an active role in the activity. There is no way we can imagine that the person in this situation is being entirely passive and is just along for the ride. In the end, *one's sense of dignity is enhanced if one manages to complete the difficult task*.

This example shows that it is possible to feel satisfaction for carrying out difficult actions that are still, at the level of atoms and molecules, completely predetermined by nature. One does not need to benefit from minor miracles in order to experience this satisfaction in the results of one's own efforts. There is a dignity associated with carrying out work and moral actions. It arises from the *control* one exerts over one's surroundings by taking action, and of producing an outcome that is consistent with one's values.

This dignity is perceived subjectively, *but it is real*. The nature of this dignity is that, as an intelligent and caring organism, one can use one's actions to control how some things happen. Although determined by nature, the actions are under an intelligent organism's control, *and the results differ from what they would be if an intelligent organism were not present.* Thus, affirming that our actions are determined by nature does not prevent us from experiencing the dignity of taking control of some of what happens in our surroundings.

TAKING RESPONSIBILITY

If our decisions are really ultimately determined by nature, it may seem that the idea of human responsibility is misguided. Many forms of responsibility are left intact, however, and these will be discussed in this section.

In the previous chapter, a special definition of the word *responsible* was given, written in quotes ("responsible"). By this definition, we are not "responsible" for those of our actions that are ultimately and completely shaped by outside causes. It was argued that we are not "responsible" for any of our actions. This follows since nature determines the outcomes of all of our decisions. *In thinking about moral responsibility, it may be useful to use this definition of "responsible."* This part of the responsibility issue will be discussed later in this chapter in the section on religious implications. One's responsibility to obey laws will also be treated later, in the next chapter.

There are other kinds of responsibility for which the definition of "responsible" given above is not relevant. There is the responsibility that involves living up to one's previous promises. These have to do with such things as responsibilities at work, marriage vows, and promises one has made to friends, relatives, and other people who belong to various organizations that one has joined. *One's actions are what count with this kind of responsibility.*

Society expects us to fulfill our commitments whether or not our actions are shaped by outside causes. There are usually some acceptable excuses for not fulfilling one's commitments, but the acceptable excuses usually do not cover many of the subtle ways in which nature can determine our actions. Instead, if one does not have an *acceptable* excuse for not meeting a commitment, one's only recourse may be to say that one is sorry and ask for forgiveness or leniency.

Making commitments usually involves only a limited responsibility. In previous centuries, captains sometimes stayed on board a sinking ship. Generals who were defeated in a crucial battle sometimes killed themselves. Debtors who could not pay their debts were thrown into prison. Nowadays, however, a leader of a corporation that loses billions of dollars because of bad management decisions may fear nothing worse than losing his or her job and being unemployed for a few months. A person who falls into excessive debt is not punished and can declare bankruptcy. This *softening* of responsibility may be society's way of dealing more realistically and humanely with people who are expected to be responsible, but who in reality are not able to control much of what happens in the real world.

What can we do in a world where nature determines our actions and society expects us to carry out our commitments? One can never be absolutely sure that it will be possible to fulfill one's promises. *One can be prudent and make commitments only if one is quite determined to carry them out and if one believes it is likely that one will actually carry them out.* Then, one rolls up one's sleeves and works hard to keep one's promises.

Will the fact that a person does not believe in Objective Free Will (**OFW**) affect the person's ability to carry out responsibilities? The person will not feel that he or she can carry out an obligation by using the kind of free will that involves minor miracles. On the other hand, a person who believes in **OFW** may hope to carry out a responsibility by making use of minor miracles, so that he or she can do what otherwise would be impossible. If minor miracles really don't happen, however, the person who relies on them is overconfident. This person may promise too much and fail. From this, it seems that a person who does not believe in **OFW** may be more prudent in accepting only *those responsibilities that he or she can fulfill.*

Does one's lack of belief in **OFW** change one's ability to fulfill promises or commitments? Presumably, one will have the same motivation to carry out one's promises whether or not one believes in **OFW**. As a result, one's success in carrying out a given promise should be about the same whether or not one believes in **OFW**. In some cases, however, the excessive *hope* held by someone who believes in minor miracles may help that person achieve something that is just barely possible. As pointed out above, however, a person who does not believe in minor miracles may be less likely to take on impossible projects than one who believes.

On the whole, it does not seem that believing or not believing in minor miracles makes much difference in one's ability to fulfill one's commitments and promises. On the grounds that *not believing in minor miracles is correct*, it seems likely that this approach leads to more success in making appropriate promises and fulfilling them. A high standard of responsibility is certainly possible for people who do not accept **OFW**.

FAILURE, REGRETS, AND TOLERANCE

One of the best results of giving up the belief in minor miracles is the effect on how we regard our past decisions. Since our decisions followed the only course that was actually possible for us at the time they were made, *there is no reason to condemn ourselves for our choices*. With things as they were when the decision was made, we could not have made different choices. Not just "what's done is done," but, *nothing else could have been done*.

Similarly, it does not make sense to flagellate ourselves for past failures, since the outcomes of all our decisions were determined by natural processes. Perhaps we could have avoided failure by working harder or having more knowledge, but these were not real options when we did what we did. To work harder, some circumstances would have had to be different than they actually were at the time that our efforts resulted in failure. To know more, we would have had to have done more thinking or to have had more experiences than we did. *The failure could only have been avoided if things had been different than they really were.*

If you have suffered feelings of guilt or regrets for past failures when you believed in **OFW**, these feelings may gradually diminish if you accept a worldview that asserts that minor miracles are impossible. A large part of the feelings of self-blame are illogical in such a view. By reflections on the implications of this worldview, your feelings may eventually attain consistency with the implications of your new beliefs.

The process by which our feelings adjust to our new beliefs about free will is called *affirmation* by the prominent contemporary British philosopher Ted Honderich. His ideas on free will resemble those expressed here. For those who would like to go deeper into this, I recommend his discussion of affirmation in his very readable book, *How Free Are You?*[4]

While rejecting **OFW** can reduce one's self-blame for bad decisions, it can also reduce one's feelings of self-praise for good decisions. *Self-admiring feelings make little sense if one believes that one's decisions were ultimately and completely shaped by nature.* It is especially true that *self-righteous feelings ought to be diminished* if one agrees that one is not "responsible" for one's morally correct decisions. On issues of morals, the rejection of **OFW** leads to a major reduction in both praise and blame. Of course, one still may be pleased or disappointed by one's actions, depending on whether or not they were consistent with one's values. One may also make plans to avoid regrettable behavior in the future.

For the same reasons that it is appropriate to experience reduced feelings of praise and blame toward *oneself* if one rejects the belief in minor miracles, it is also appropriate to reduce one's feelings of praise and blame toward *others*. The rejection of the **OFW** belief is likely to make one more tolerant of other people's behavior. Ultimately, their choices and actions are caused by circumstances arising outside themselves. *Realizing this will tend to make one more accepting of their choices and actions.*

Although we may become more tolerant of others, we still can try to modify their actions by approving or discouraging certain kinds of behavior. It is one thing to be self-righteously condemning of others, it is another to try to improve other people's actions. Our approval or disapproval may be the deciding circumstance that will influence what others will do. *We are part of the outside world that shapes other people's behavior.* We can be tolerant without being fatalistic about the behavior of others.

RELIGIOUS IMPLICATIONS OF THE NATURAL SOUL

There is a sort of *central constellation of common Christian concepts* (CCCCC) about reward and punishment that is contrary to some of what has been said here about the natural soul. Many Christians may not hold all of these beliefs, but they are common enough that they are worth discussing. Most of these beliefs were held by the Zoroastrians and are also shared by some Jews and Muslims.

The following beliefs are part of the CCCCC. A personal God exists. This God is good and just. God expects people to behave in certain ways. People have a free will of the type that was called Objective Free Will (**OFW**) in this and preceding chapters. That is, in making moral decisions, people are believed to have the ability to choose either a better or a worse option. The choice is not completely determined by nature. People are believed to have some kind of immortality. After death, God rewards or punishes people according to their actions and/or their faith. In this picture, earth is a sort of testing ground for people and God is the ultimate source of justice. According to CCCCC, it makes sense for God to condemn people for making poor decisions.

If one believes in the *natural soul* rather than the *supernatural soul*, the CCCCC picture is not very reasonable. The natural soul is not immortal, so it makes little sense to maintain that God grants justice after death. Also, the natural soul does not have the power to work minor miracles that makes **OFW** possible. So, one would not expect a just God to punish humans for making the only choices that were possible for them to make. Thus, it makes no sense for God to punish people for their good or bad deeds. Similarly, a person who did not have faith *could not have chosen to have* *faith.* If we do not have **OFW**, it does not seem to make sense to punish a person for lack of faith, no matter how "faith" is defined.

In summary, if one does not believe in **OFW**, it does not make sense to believe that God should punish people for their deeds or their lack of faith. If God is both just and loving, one might reason that God could *reward everyone*, since *there is no fair basis for discriminating between people based on their deeds or faith.* Still, if the idea of a natural soul is true, there seems to be no immortality, so one can ask when this reward would be granted.

Accepting the idea of a natural soul destroys the worldview in which one is motivated to do good and avoid evil because of the desire for heavenly rewards and fear of punishment in hell. There is still the motive of *doing good for its own sake*, because it has good effects and is the better thing to do. One can still be unselfish and consider the welfare of others as just as important as one's own. *There is also the possibility of doing good because of love of God.* This reason requires a belief in a personal God, which is discussed in the next section.

WHY WE TEND TO CREATE A PERSONAL GOD

Almost all of the gods in the world's religions are believed to be personal gods, who have mental activities and memories, and feel emotions. For humans, being a person is based on a physical structure: the body, especially the brain. This suggests that gods have brains similar to human brains. But many gods, including the Christian God, are supposed to be spirits. A spirit, presumably, would not have the extremely large number of precisely organized components that make consciousness and other mental qualities possible for humans. This suggests that *there could not be such a thing as a purely spiritual person.* One seems to need a complicated structure made of huge numbers of parts in order to be a person.

Why *do* we believe in a personal God? In our ordinary lives, we do not have direct evidence that there is a personal God. The universe does not seem to be concerned about the well-being of persons. Only an *incredibly tiny fraction* of the matter in the universe is tied up in the bodies of persons. This does not suggest that the universe was created by one Great Person so that ordinary persons would have a home. It appears that the universe functions by impersonal physical laws. As mentioned earlier, even the origin of the universe may have been a "free lunch": supernatural intervention may not have been needed.

Many people believe in God because of the Bible. In previous chapters, the discussions supported the view that the Bible has a natural origin. If this is true, the Bible should be viewed primarily as the concentrated wisdom of spiritual leaders who lived about two thousand or more years ago. If the Bible has a natural origin, the evidence for God in the Bible should not be trusted much more than the evidence for God or gods in the holy texts of other religions. Few people believe in *all* of the gods of all the world's religions!

We have little or no direct evidence for a personal God. Similarly, we have no convincing and direct evidence for the existence of intelligent aliens in space, but many people believe in them. In the opinion of some scientists, intelligent aliens seem to be likely to exist. In this case, however, the lack of hard evidence for aliens from space is sometimes interpreted as indicating that the aliens *don't want to be detected*. If there is a personal God, perhaps He, She, or It does not want to be detected.

It is sometimes argued that we have a *hidden God* because a truly convincing revelation of the existence of God would undermine the process of testing the moral quality of humans on Earth. In the previous section we saw that the belief that Earth is a moral testing ground does not make much sense unless we believe in minor miracles. Assuming that minor miracles do not exist, then this explanation of why we have a hidden God is not believable.

Let's consider why there are *persons* on Earth. This has some bearing on the idea of a personal God. We have already discussed reasons why persons exist. Self-moving organisms, or animals, need nervous systems to control their movements. Nervous systems help animals select actions that improve their chances of survival. Emotions develop as a means of influencing decisions in ways that satisfy the basic needs related to survival. Advanced nervous systems include a brain that is capable of constructing representations of the outside world. This gives an animal the ability to make good choices, leading to actions that enhance its biological fitness. A special organization of part of the brain handles the really complicated mental tasks that require use of all the animal's senses and previous knowledge. *This highly organized system leads to consciousness.*

The automatic process of developing successful survival strategies in animals led to consciousness and emotions, and to all the characteristics that make what we call *persons*. Our existence as persons is directly linked to the evolutionary struggle for survival. This struggle is relevant only for *mortal* organisms, but, of course, all organisms are mortal. The conclusion of all this is that *the development of persons on Earth is closely tied to the fact that these persons are mortal organisms*. Mortality drove the process of evolution to produce persons who have mental abilities well-adapted for survival.

The reasons why *we* are persons could not be the reasons why a supreme being would be a person. If the supreme being resembles the Christian and Jewish God, the supreme being would not be mortal or in competition with other potential supreme beings. Such a supreme being would presumably not have an evolutionary history that produced a nervous system, and so on. *Without a nervous system, there is no reason why a supreme being would be conscious.*

Our kind of supreme being would have no need for reproduction. Consequently, it is more reasonable to refer to *It*, not to Him or Her. It is only because God has been imagined to be a person like us that we refer to God as if It were a sexual being. A supreme being would not need emotions, for It would not need a built-in motivational system to help It survive. *Why should we believe such an entity would be a person?*

The fundamental reason why we believe that God is a person is that *our predecessors created God in our own image*. The only way that ancient religious thinkers could understand the existence of *purpose* outside of human activities was by hypothesizing a God with properties like those of a person. Until not very long ago, the impressive adaptations of animals and plants were accepted as evidence of purpose in nature, rather than as a natural outcome of evolutionary processes. The wonderful ways in which organisms are adapted to their environment was thought to arise from their being designed by *an intelligent creator, who was obviously a very wise person.* It was not a big step to conclude that *the fundamental cause of our being or existence was a supernatural act of will by this enormously powerful person.* It is especially ironic that, after creating God in our own image, we turn this around in a self-congratulatory statement, saying that *we are made in the image of God.*

It is so common to believe that there is a personal God that we usually do not ask why we believe this. As noted above, the common reasons for believing in a personal God are not convincing when considered in light of reason and scientific knowledge. The Bible is solid evidence that *ancient Hebrews* believed in a personal God, but it is not such an infallible document that *we* should trust it on such a fundamental issue. There is no hard evidence for the existence of a personal God. *Persons* result from evolutionary processes in material systems. These processes seem to have nothing to do with spiritual entities. In summary, there is no convincing reason to believe in a personal God.

There are some radical consequences if we accept the reasonable conclusion that the fundamental cause of our being is not a person. *There is not a personal God who can be flattered by our praise or pleased by our obedience or our faith. It is up to us to choose our values and virtues and to find "meaning" in life.* We are not bound to follow religious teachings that are based on a belief in a personal God. On the other hand, we do not need to alter our moral principles just because they are no longer based on a religious creed.

HAVING VALUES AND LIVING BY THEM

If there is no personal God and no punishment or reward after death, there is no motivation to be morally good for love of God or fear of hell. Still, there are reasons for doing good. Much of morality is *a useful standard of behavior in any civilized society*. Much of what is regarded as good behavior in one civilized country is also respected in other civilized countries. We saw that most of the Ten Commandments resemble rules of the more ancient Egyptian religion. As another example, consider the Far Eastern cultures such as those of China and Japan. Although their principal religions are very different from Christianity, things are not entirely different there. They also have rules against murder and theft, and they share many values with us, such as honesty, hard work, and devotion to family life. Many rules of good behavior are widely held, but we tend to emphasize the differences and ignore the similarities between the rules of various nations. For many of us, this tendency is strengthened by the belief that *our own rules were set down by God*.

Although people may do good deeds to please God or strive for otherworldly goals, the approach advocated here is that *we try to live in harmony with our values and the reality that surrounds us*. By values, I mean our beliefs about how one should act. In this way of thinking, a major motivation for good behavior comes from thinking about what the likely consequences of one's behavior will be. Good actions are aimed at achieving health and happiness for oneself and for others. If one's attempts are successful, one will be rewarded by personal health and happiness, as well as satisfaction that one has assisted others in reaching such goals. As with planting fruit trees, there are natural rewards if the behavior is successful, but the goals may not be attained.

In the book *Honest to Jesus*, Robert Funk calls the natural rewards of good behavior *intrinsic rewards*.[5] According to Funk, Jesus taught that doing good would yield intrinsic rewards. Very early in Christian history, however, the developing religion adopted the mythical ideas of heaven, hell, and the final judgment. These mythical motivations for doing good are the *extrinsic rewards* that are so prevalent in traditional Christian teaching. According to Funk, *the promises of extrinsic rewards are a deviation from the actual teachings of Jesus*.

Some readers may suspect that it is sometimes not possible to act according to one's values if one's actions are "shaped by the universe." That is, if an action is *the only action that could have resulted* from the physical causes that preceded it, how can we expect the action to turn out to be consistent with our values? The answer to this question has a "good news" and a "bad news" part.

The "good news" part is that our values are encoded in connections of neurons in our brain. The same neural connections that determine that we *hold* certain values *steer our actions toward behavior that is consistent with those values*. From the high-level perspective of complete mental processes, our values serve as one of the causes of our behavior. On the *microscopic* level, the numerous neural connections that determine that we are aware of holding certain values also tend to cause us to make decisions that uphold these values.

The "bad news" part of the answer is *that we sometimes take actions that are not consistent with our values*. Sometimes we make decisions so quickly that we overlook likely consequences that are opposed to our values. This lack of reflection can cause a person to do something that seems inconsistent with his or her character. Another "bad news" fact is that our values are not the only motivations that we have. For example, a person may make a bad decision when he or she is angry. This decision may be motivated primarily by the emotion rather than the individual's system of values. Unless one's values are strongly in control, one may often perform deeds that are not consistent with one's values.

The "bad news" may suggest that there is something wrong with the view that one's behavior is determined by physical causes. One may wish to live in a world where one's values always control one's actions. For example, if Objective Free Will were true, one *might* always make the proper decision. We have, however, already argued that **OFW** is only a myth. In any case, even the most supernatural-minded supporters of traditional Christianity will admit that sin is rampant within the world. *Even the traditional approach based on supernatural occurrences cannot guarantee that our behavior will always be good.* There is no reasonable picture of our nature in which our values always control our actions.

Christianity has often had difficulty in accounting for the *existence of evil* in the world. The existence of evil is not hard to explain, however, if one accepts the worldview in which a personal God is not present. In this view, the nonhuman world neither supports nor opposes evil. An evil act may be carried out by a person who *lacks values entirely, acts without sufficient reflection,* or *is dominated by motivations other than the desire to live according to his or her values.* All of these reasons are entirely natural.

Although the *nonhuman world* is neutral toward evil, *people* may have values and may be determined to live according to those values. People are intelligent enough that quite abstract values may control their behavior. We can foresee the consequences of our actions and this can rule our actions. We are capable of "putting ourselves into the other fellow's shoes" so that we become willing to work for the health and happiness of other people. As animals of such an intelligent and civilized nature, the "law of the jungle" is normally not the controlling rule.

The worldview supported here *encourages the individual to take full ownership of his or her life, not to see life as a franchise bestowed by a God who retains rights over the body and the soul.* This view also suggests that we start over in forming our goals and values. Still, the process of reform may not lead to major changes. Concerning oneself, one might aim for health and happiness, using such strategies as avoiding harmful drugs, while employing education, hard work, and careful planning to promote one's goals. Concerning others, one may help family members, friends, and even strangers when it is appropriate. To avoid harming others, one might aim for fairness, truthfulness, tolerance, and obedience to just laws. One may also try to improve conditions for others by supporting education, family planning, a healthy environment, and uniform and universal assistance for those who need it.

The list of goals and values given above is quite subjective. Readers may have different ideas about the most important values. The main goal of this section is not to give a list of worthy values. It is to emphasize that *a worldview that lacks a personal God and extrinsic rewards can still include respect for good behavior and for most conventional values.* This view supports *doing good for its own sake,* not for the hope of heavenly rewards.

NOTES

1. Elisabeth Kubler-Ross, *Questions and Answers on Death and Dying* (New York: Macmillan, 1974), p. 159.

2. Ibid., p. 162.

3. Robert Kavanaugh, *Facing Death* (Los Angeles: Nash Publishing Co., 1972), p. 221.

4. Ted Honderich, *How Free Are You? The Determinism Problem* (Oxford: Oxford University Press, 1993), chap. 9.

5. Robert W. Funk, *Honest to Jesus: Jesus for a New Millennium* (San Francisco: HarperSanFrancisco, Polebridge Press Book, 1996), p. 312.

16

SOCIETY AND OUR NATURE

MANY OF OUR BELIEFS ABOUT social issues arose from ancient ideas about supernatural souls and a personal God. Thus, the conclusions of the previous chapters affect such emotionally charged topics as the value of human life, abortion, birth control, criminal justice, and the future of religion.

SENSIBLE REVERENCE FOR LIFE AND CONSCIOUSNESS

Although science gives us a new way to understand the world, life, and consciousness, it does not take away our appreciation of these things. Science helps us comprehend the awesome nature of our physical surroundings, the amazing reality of our own awareness, and the wonder of life.

Unlike the physical world, *life must be protected if it is to survive*. The many species of plants and animals on Earth are probably not accurately duplicated near any of the other billions of stars in our galaxy. As living beings ourselves, we tend to feel that all of these species are valuable and should be preserved.

There are practical arguments for maintaining all of the species of life that occur on Earth. Some species possess unique biochemicals that may help us avoid diseases. Plants and animals may have unknown beneficial effects that we do not understand at present. Lost species may be lost options for future benefits. It is impossible to know just how important such concerns are, but it is only prudent to follow the safer path and preserve rare species, even in the face of some economic losses.

Conscious life is even more rare and valuable than life without awareness. *At present, we do not know whether conscious life is present anywhere else in the universe!* We are conscious and value our own lives. By analogy with our own feelings, and since valuing one's life has survival value, *we expect that other people also value their own lives*. Most people feel that other human beings have some value, at least to those persons themselves. Except in war or unusual circumstances, most of us are willing to respect other people's lives. Normally, in a sort of reciprocal and beneficial relationship, we allow others to live and demand that others allow us to live.

The value of one's life is based primarily on one's own desire to continue living and, to a lesser extent, on the value placed on one's continued existence by other people. Although we often have an exaggerated sense of our own importance, *society can get along almost equally well with or without one's continued existence*. If one accepts this picture, it is reasonable to believe that people should have the right to terminate their own lives in extreme situations where continued life is a painful burden. Under extreme circumstances, *suicide* may be a reasonable option.

Suicide is strongly condemned in some religious traditions. From such a perspective, one does not possess full ownership of one's life. Life may be regarded as a supernatural franchise bestowed by a God who retains rights over the body and the soul. On the other hand, if matter, life, and consciousness arose by *purely natural processes*, there is no reason to accept a prohibition against suicide based on our supposedly supernatural origins. If a personal God is a mythical creation of human imaginations, then it makes no sense to condemn suicide on the grounds that God has the sole right to control the time of one's death.

If a person's main value arises from his or her conscious desire to continue living, then some religious arguments against *abortion* have missed the point. From the discussion of consciousness in previous chapters, it makes no sense to believe that an embryo is conscious before it has developed a nervous system. At such a point the embryo is not capable of mental activity or any kind of awareness. So, unless one believes that consciousness can arise from supernatural processes outside the brain, one is led to believe that *destroying a fetus before it has developed a functioning brain is not anything like killing a person capable of consciousness.*

SEEKING JUSTICE BY NATURAL MEANS

> *A man can no more be blamed for having a weak "moral fibre," than he can for having a weak kidney.*
> —Euan Squires, *Conscious Mind in the Physical World*[1]

Our traditional beliefs about free will give us exaggerated feelings of responsibility for our own actions. The mistaken idea of Objective Free Will (**OFW**) leads some people to hold that criminals deserve harsh punishment because they could have avoided making their evil decisions. Other traditional beliefs suggest that we can leave justice to God, and that we will be rewarded or punished for our deeds after we die. This belief collapses if we stop believing in a personal God and immortality.

Those who don't believe in immortality or a personal God do not expect justice to be delivered in a future life. If justice is going to be delivered, it has to be done here on Earth. These people believe that our present life is the real thing. It is not a mere test of one's worthiness. Life is of importance in itself, not because it determines one's status in a future life.

If we deny **OFW**, we acknowledge that a criminal's actions were unavoidable, given the detailed physical situation at the time of the crime. *It is not fair to punish a criminal severely for carrying out an unavoidable action.* An extreme response, based on fairness, would be to avoid punishing anyone. In practice, this is not acceptable, since society needs to deter crime. But isn't it unfair to punish people for actions that really were unavoidable? We may be willing to tolerate this unfairness in order to have a more orderly society. In any case, harsh punishments for crime are unfair. It is better to retrain or restrain criminals, rather than torture or execute them. Hopefully, society will find ways to prevent crime without punishing perpetrators *after the crimes have been done.* In itself, the fact that *there are criminals in spite of deterrence* implies that deterrence has limited effectiveness.

Why is the discussion of crime devoted primarily to the criminal? If the crime was really unavoidable under the circumstances, isn't justice served more effectively by helping crime's *victims*? If we admit that a society produces the situations that lead to crime, it seems especially appropriate for governments to compensate crime's victims. This might involve medical care, counseling, restitution of property, or compensatory payments, depending on the circumstances. This could help the victims much more than, for example, letting them view the conviction or even the electrocution of the criminal.

SEEKING A HEALTHY WORLD

All life as we know it is confined within a thin shell on the surface of our planet. This shell contains the most valuable property in the known universe. To protect life's future, we must maintain this region's precious ability to continue to support life. This requires that we avoid three things: *destruction of the environment, nuclear war, and uncontrolled population growth.* In my opinion, these issues carry a greater moral weight than the typical issues stressed by the world's religions.

Obviously, protecting the environment and avoiding nuclear war are desirable goals. Sometimes shortsighted thinking encourages people to disregard these all-important long-term goals. It is important to keep concerns about the environment and nuclear war in mind whenever we vote or participate in other political processes.

Some readers may not feel that controlling population growth is as important as protecting the environment and avoiding nuclear war. The trouble with unlimited population growth is that the environment can handle only a limited number of people and the wastes that are directly and indirectly produced by these people. Severe overpopulation makes it almost impossible to maintain forests and clean rivers and lakes. The pollution produced by a dense human population is sometimes reduced if the people are very poor, since they use little fuel or materials. But *unlimited* population growth guarantees that desperately poor people will eventually overtax the environment, or that the population will eventually be controlled by epidemics, mass starvation, and wars. None of these results are acceptable.

The idea that human conception involves a sacred or supernatural process has been used by certain religious leaders to argue that some or all forms of birth control are sinful. A skeptic may suspect that some religions desire unlimited growth, and that restricting birth control helps these religions reach this goal. In the worldview supported here, human conception is an entirely natural process. It is reasonable for potential parents to control their family's size by preventing conception when it seems appropriate. In this view it seems irresponsible to conceive children when they are not wanted or cannot be supported.

RELIGION WITHOUT SUPERNATURALISM

One goal of this book has been to present a picture of reality that contradicts *supernaturalism*, the belief in supernatural events and entities. It appears that there is no convincing evidence supporting supernaturalism, and no evidence favoring one form of supernaturalism over any other.

Some of the most successful religious groups pull their members deeper and deeper into supernaturalism. Since almost all religions favor some form of supernaturalism, support for religions may diminish if people conclude that supernaturalism is mistaken. Even in a largely secular society, however, religion serves many functions, such as the following:

- Religion educates people in ethics and morality.
- Religion connects people to their history, ancestors, and traditions.
- Religion helps people face problems by emphasizing what really matters.
- Religion binds people to one another in a way that is often rare outside religion.
- Religion provides ceremonies to mark life's major transitions.
- Religion provides people with inspiring and consoling music and rituals.

In the absence of religion, these functions can be provided by the family and various other social groups. Practicing a religion, however, is a convenient way for people to fulfill their needs for all of these services. As attitudes shift away from supernaturalism, some religions are becoming more secular and are concentrating on providing the services listed above. There is a trend in this direction in some religious groups whose members have higher levels of education. This trend makes it easier for more educated and intelligent members to enter their churches without "checking their brains at the door."

Our churches do not need to be the only teachers of ethics and morality. Many nations with entirely different religious beliefs have generally high standards of ethics and morality. In chapter 3 we saw that, long before Judaism and Christianity, the ancient Egyptians had many of our most cherished ethical and moral standards. Whether or not the ancient Egyptians were the source of these ideas in Judaism and Christianity, it appears that reasonable ethics and morals can be relatively independent of the details of one's supernatural beliefs. This suggests that nondenominational ethics and morality could be taught in public schools. Unlike religious education, it could be stressed that ethical and moral principles are not absolute truths, but examples of rules of behavior that previous societies have found are helpful standards.

Unlike members of many other religions, many Christians believe that their myths are literally true. Chapters 3 and 4 discussed how the supernatural events in the Bible seem to be mythical. If church members conclude that their supernatural beliefs are mythical, the churches can adapt to this. Churches could admit that their myths are not literally true without completely destroying their appeal. In practice, the clergy are sometimes more aware of the mythical nature of certain church teachings than the members. This leaves priests and ministers in a very difficult situation in dealing with members who are blessed with excessive faith.

In the previous chapter it was emphasized that one does not need to believe in souls and live by myths in order to live a decent life. A widespread acceptance of the views expressed in this book would not harm society. For most of us, our best chances for success and happiness depend on holding beliefs that are in harmony with modern knowledge.

CLOSING COMMENTS

Earlier chapters argued against the idea that consciousness and our other mental activities transcend the laws of science. Science finds no convincing evidence of such transcendence. Consciousness has evolved to enable animals to be aware of their surroundings, so they can select actions that help them survive and reproduce. As such, it is a very specialized property of a miniscule portion of all the matter in the universe. To suggest that consciousness depends on special laws of nature or supernatural effects brings to mind past claims that Earth is the center of the universe or that life is impossible without supernatural intervention.

Previous scientific research has shown that life is entirely the result of material interactions. Currently, science is finding that our awareness results from purely physical processes. As usual, however, our imaginations are too limited to do justice to the future progress of science. Although science tries to minimize its dependence on "faith," perhaps the best use of faith is to bet that science will continue to make progress in areas that are changing rapidly at present. Doing this, I predict that future generations will learn how the "subjective" aspects of consciousness result from natural brain processes.

Even though the common belief in souls is probably mistaken, it is understandable that most people believe souls give people abilities that seem beyond the reach of material processes. People will probably change their opinions about souls in the not too distant future. The public will be better educated and most will not believe that the Bible is literally true. As the natural bases for our existence become better understood, people will lose faith in "explanations" based on supernaturalism. The idea of a future eternal life will seem cultlike and less likely. As confidence in immortality diminishes, people will seek fulfillment in their natural lives, as perhaps the majority already do.

Partly because it is *not* so evident that it is based on supernaturalism, what I have called Objective Free Will (**OFW**) is likely to be widely held well into the future. It is now held so deeply that many readers may feel that it is perverse to deny it. By supporting a rigid view of moral responsibility, **OFW** *seems* to be an important element in maintaining an orderly society. I maintain that we can get along very well without it, and that believing in it amounts to assuming that minor miracles are common occurrences. In conclusion, I am now in a position to say, with all due humility, that it would require one or more minor miracles for me to be completely mistaken about this issue.

NOTE

1. Euan Squires, *Conscious Mind in the Physical World* (Bristol, England: A. Hilger, 1990), p. 235.

Appendix I

HOW THE QUANTUM
ENTERED MODERN SCIENCE

THE IDEA OF THE QUANTUM developed from discoveries made in physics early in the twentieth century. Certain predictions of classical physics did not agree with experiments or were unreasonable.

Classical physics failed to predict the correct density of the energy of heat, light, and ultraviolet radiation inside very hot cavities, such as an intensely hot furnace. According to classical physics, the total energy density of such radiation would be *infinite*. What was observed in experiments was a sensible, *finite* density of radiation. The relative amounts of energy measured at various wavelengths also disagreed with what was predicted by classical physics.

A solution to these discrepancies was found by the German physicist Max Planck (1858–1947) in 1900. To predict the total electromagnetic radiation in a hot cavity, one needs to add the contributions from all frequencies of the electromagnetic spectrum. In the calculations made before Planck's work, the amount of energy at each frequency was allowed to be zero or *any* positive value.

Planck tried another approach. He introduced a new quantity to physics, denoted by h. Later, it was called Planck's constant. He assumed that at each frequency, f, the cavity would contain an amount of energy equal to zero or some *whole number* times h times f. He allowed no *fractional* multiples of h times f. In other words, he allowed the energy to occur only as a certain number of packets, or *quanta*, each of size h × f.

Using an appropriate value for h, Planck found that his calculation gave correct answers for the total radiation within a cavity. It also predicted the actual amount of radiation that was observed in each frequency interval. Part of the significance of Planck's work was found later. In 1905 Einstein looked at the problem of radiation in cavities and concluded that the radiation behaves as if it is a gas of particles, each carrying an energy equal to h × f. *These particles are now called photons*. Photons are particles that carry electromagnetic energy. Beams of infrared radiation and light are streams of photons.

The accepted scientific opinion about the nature of light at the start of the twentieth century was that light consists of *waves*, not particles. Many experiments supported this belief. After the general acceptance of the particle nature of light, the picture was so confusing that, "One wit remarked that on three alternate days of the week the experiments supported the photon theory, whereas on three other days the evidence supported the wave theory; thus it was necessary to use the seventh day of the week to pray for divine guidance!"[1] Even today we still believe that light can exhibit either wave or particle properties, depending on the nature of the experiment that is performed with the light.

Before long, other quantum effects were found. (By quantum effects, I mean that certain basic processes in physical systems appear to involve standard-sized bundles or *quanta*.) In chapter 5 we saw that Rutherford showed that the atom consists of electrons in orbit around a nucleus. Classical physics, however, was not able to explain some properties of atoms. It was not possible to understand how the electrons could stay in stable orbits.

Due to their constantly changing directions in their orbits, the orbiting electrons should have emitted so much light and electromagnetic radiation that they would rapidly lose their energy and spiral into the nucleus. According to classical physics, this was expected to happen in only 0.01 *millionths* of a second. In reality, it is possible for atomic electrons to continue orbiting the nucleus in their lowest energy levels for indefinite time intervals.

Another thing that classical physics did not explain properly was that the radiation from the atoms in the processes described above would be produced at all frequencies. In reality, the emission of light from tubes of electrically excited gases occurs at certain specific frequencies. Tubes filled with different gases, such as neon, argon, or krypton, emit light at different frequencies. Consequently, they have different colors.

Hydrogen gas can emit light at a number of different frequencies. The different frequencies are related to each other by fairly simple mathematical relationships. These relationships involve small whole numbers such as two, three, and five. This suggests that some kind of quanta are involved.

Niels Bohr (1885–1962), a young Danish physicist working at Rutherford's laboratory, proposed a simple theoretical theory of the atom in 1913. Bohr proposed that the *angular momentum* of electrons orbiting the atom's nucleus can only have certain values. For a circular orbit, the angular momentum of an electron is the product of the electron's mass times its velocity times its distance from the nucleus. According to Bohr, the angular momentum could only be equal to a whole number, n, times Planck's constant divided by $2 \times \pi$. (The quantity π is 3.14159 . . . , the ratio of the circumference to the diameter of a circle.)

The Bohr theory of the atom allowed electrons to circle the nucleus only in certain specific orbits, corresponding to values of n, such as one, three, or seven. Different values of n correspond to different electron energies as well as to different angular momenta. According to the theory, atoms emit light when an electron jumps from one orbit to another orbit of lower energy. The emitted photon carries an energy equal to the difference between the electron's energy in the initial and the final orbit.

Bohr's simple theory predicted the correct values of the many different wavelengths of light produced by hydrogen. His theory gave a reasonable value for the size of the hydrogen atom. In his theory the size of the atom is the diameter of the electron's orbit. An interesting feature of atoms in Bohr's theory is that, like the random decay of nuclei, the electron could change its orbit and emit a photon at some unpredictable time. Thus, the idea of *unavoidable randomness* was already present in quantum theory at this point.

Bohr's theory of the atom uses some arbitrary assumptions. For example, only specific orbits are allowed and electrons in these orbits do not radiate their energy away as predicted by classical physics. As quantum theory became more advanced, it was found that it was not necessary to use such arbitrary assumptions in order to calculate the atom's properties. It became possible to *obtain* these properties as *results* of the more advanced quantum theory.

NOTE

1. Nathan Spielberg and Bryon D. Anderson, *Seven Ideas that Shook the Universe*, 2d ed. (New York: John Wiley & Sons, 1995), p. 276.

Appendix II

WAVES, UNCERTAINTIES, AND RANDOMNESS

T HE FRENCH PHYSICIST LOUIS DE BROGLIE (1892–1987) made an extremely important advance in quantum theory in 1923. He noted that *photons* can exhibit the properties of particles or of waves in different kinds of experiments. Although photons have no mass, he proposed that *particles with mass* can also exhibit wave-like properties. *By a rather peculiar derivation, he argued that a massive particle's wavelength is equal to Planck's constant divided by the particle's momentum.* In general, a *wavelength* is the distance from a point on one wave to the similar location on the next wave, such as the distance from the crest of one wave to the crest of the next wave. The particle's momentum is the product of its mass and its velocity.

Although de Broglie's proposal was radical, the formula for the wavelength had some things in its favor. For ordinary-sized objects moving at reasonable speeds, the *de Broglie wavelength* is so extremely short that it has no detectable consequences. This was good, since no effects of the waves had ever been detected in objects that were not microscopically small.

Another nice feature of the proposed de Broglie wavelength was that it could be related to Bohr's theory of the atom. De Broglie used his formula to calculate the wavelength of electrons in orbit around an atom's nucleus. He found that Bohr's rule for the allowed values of the angular momentum of atomic electrons amounts to having a whole number of de Broglie wavelengths in an electron's orbit. This allows the electron's waves to stay in phase with each other over successive orbits. This seemed as if it might allow an electron to stay in a fixed atomic orbit without emitting light or other radiation.

The real support for the wavelike nature of massive particles came in 1927, when beams of electrons were scattered off crystals. The scattered electrons showed patterns that appeared to be produced by *waves striking the crystals. Experiments also verified de Broglie's formula for the wavelengths of the particles.* The wavelike nature of electron beams is now used in electron microscopes, allowing objects to be observed that are too small to be seen with optical microscopes. *It is believed that all objects have a wavelike nature, but this is detectable only in objects about as small as or smaller than atoms.*

In 1926 Erwin Schrödinger (1887–1961), an Austrian physicist, learned how to predict how particles behave when they act like waves. Using *Schrödinger's equation,* one could understand the structure of *atoms* in a much more complete and fundamental way than by the use of Bohr's theory. In classical physics, the initial conditions and Newton's laws were used to predict the future behavior of a physical system. In quantum theory, however, *boundary conditions* and Schrödinger's equation were used to find the expected behavior of such microscopic things as elementary particles.

Quantum theory gave physics a new and accurate description of how extremely tiny systems operate. Quantum theory is able to predict many properties of atoms, molecules, solids, and particle collisions that were impossible to predict using classical physics. There are major differences between how classical physics and quantum theory make predictions. When one "solves" Schrödinger's equation, what one gets is not the location or velocity of a particle, but something denoted by the Greek letter psi (ψ), called the *wave function.*

For a specified time and location, the value of the wave function is a *complex number*. Since complex numbers have *both* an amplitude and a phase, they are especially suitable for describing a wave. The *amplitude* represents the size or strength of the wave. When two waves overlap, the relative *phases* of the two waves determine whether the size of the combined wave is the sum or the difference of the amplitudes of the waves, or some value in between these two extremes.

The wave function contains all the information that there is to know about the system that is being studied. Just what is this wave? The answer was given in 1926 by the German physicist Max Born (1882–1970). *The square of the amplitude of a particle's wave function ($|\psi|^2$) at some location is a number that is positive or zero. This number gives the probability that the particle is located at that position.*

Let's discuss what the previous sentence means. If I know ψ, I can compute $|\psi|^2$ at some location. Then I can use $|\psi|^2$ to compute the likelihood that the particle is in some specified volume surrounding the location. Almost always, however, $|\psi|^2$ is positive over some extended volume of space, so that *I don't know exactly where the particle is*. I only know that the particle must be *some place where $|\psi|^2$ is positive*, not where $|\psi|^2$ is equal to zero. The particle's most likely location is where $|\psi|^2$ has its largest value.

In classical physics we could calculate the precise position of a particle. In quantum theory, however, the predicted location is almost always known only to be within some volume of space where $|\psi|^2$ is positive. The particle will be found at some random and unpredictable location within this volume. We might have expected that something like this would happen when we learned that particles can act as if they are waves. Waves, except for solitary "spikes" that exist only at a point, extend over some region. Since the waves are related to the possible locations of a particle, quantum theory gives answers that allow particles to be located at a range of locations.

A rule about the uncertainty of a particle's position and momentum was given by the German physicist Werner Heisenberg

(1901–1976) in 1927. The Heisenberg uncertainty principle states that *the uncertainty in a particle's position, multiplied by the uncertainty in its momentum, is at least as large as Planck's constant divided by 2 × π*. (Many rules of quantum theory involve Planck's constant.)

Another uncertainty principle is the same as that given above except that energy and time replace position and momentum in the rule. *This allows violations of the conservation of energy to occur if they stop after extremely short time intervals*. This was mentioned in connection with Yukawa's theory of mesons in chapter 5. The uncertainties we are discussing are not merely due to the limitations of technology or the interactions of equipment with the systems being studied. They are unavoidable and *they even affect physical processes in which no measurements are made. They seem to be part of the intrinsic nature of matter*.

Quantum theory teaches us that microscopic processes involve a great deal of unavoidable randomness. Uncertainties in particle positions, momenta, and energies are usually present, and atoms and atomic nuclei can emit radiation at *times* that are impossible to predict. For atom-sized objects, *quantum theory replaces the rigid determinism of classical physics with random behavior*. Einstein had great difficulty in accepting this fact, but the results of many experiments have made most physicists stop believing in determinism in the microscopic world.[1]

Although the behavior of a particular particle can't be predicted in detail, quantum theory *does* give *precise probabilities* for different possible behaviors by a particle. If one repeats the same kind of process very many times, the statistical results will agree very accurately with what is predicted by $|\psi|^2$. For example, the rates at which particles occur at some location will, except for statistical fluctuations, be proportional to $|\psi|^2$ at the location. Since $|\psi|^2$ contains predictions for very many locations, *the quantum theory does predict a great deal about how a microscopic system will behave*. Thus, the randomness in quantum theory is still orderly in that *it follows precise statistical rules*.

NOTE

1. Andrew Whitaker, *Einstein, Bohr and the Quantum Dilemma* (Cambridge: Cambridge University Press, 1996), pp. 238, 239.

GLOSSARY

Amino acids. Biochemical compounds with certain chemical properties that allow them to link together to form **proteins**.

Antimatter. Matter made entirely of **antiparticles**. If brought in contact with ordinary matter, a colossal explosion would occur!

Antiparticle. An **elementary particle** that has some properties, such as mass, that are identical with those of an ordinary particle, but for which other properties, such as **electric charge**, are the opposite of the ordinary particle. Particles and antiparticles can "annihilate" each other, changing into other kinds of particles.

ATP (Adenosine triphosphate). A biochemical constructed by plants and animals to store energy. When plants acquire energy from sunlight or when animals obtain energy from food molecules, the energy is used to construct ATP. The ATP is then used as the energy source in most cellular processes.

Attached value. As used here concerning the development of personal likes and dislikes, an automatic assignment of merit to a situation based on a person's previous experiences. For example, disliking being in a dentist's chair.

Autonomic nervous system. A major part of the nervous system that is concerned with "housekeeping details." By this system, the lower part of the brain controls blood pressure, glandular activity, and the functioning of internal organs such as the digestive system, the heart, and the lungs.

Axon. A long projection of a **neuron** that carries neural impulses from the body of the neuron to the **synapses**, where the signal can be transmitted to other neurons.

Big bang. The theory, supported by much evidence, that the universe originated in an explosion of matter at extremely high temperature and density.

Brainstem. Lower parts of the brain, including the upper spinal cord, the cerebellum, the medulla, and the pons.

Cartesian dualism. See **Dualism**.

Causal reductionism. The view that the behavior of the parts of a system (ultimately, the parts studied by subatomic physics) completely determines the behavior of the system.

Central constellation of common Christian concepts (CCCCC). As used here, CCCCC is a linked set of beliefs, perhaps of Persian origin, that lead one to conclude that people deserve reward or punishment after death. The beliefs include faith in one supreme, good, just, and personal God, who expects us to behave in certain ways. CCCCC also includes beliefs in immortality and a strong form of **free will**, called **Objective Free Will**.

Cerebral cortex. The large, folded, surface layers of the cerebrum, the upper part of the brain. The cortex is heavily involved in conscious activities.

Chaos theory. A theory that tries to find simple ways of describing the behavior of some **deterministic** physical systems that have erratic behavior. The behavior of such systems is effectively impossible to predict far in advance. Earth's weather is an example of such a chaotic system.

Christ cult. A subgroup of early followers of Jesus who emphasized his **Resurrection** and his role as a redeemer, but not his life and sayings.

Classical physics. The parts of physics that were well established by about 1900. It excludes relativity and **quantum theory**.

Conservation law. Any of a number of laws of physics or chemistry that specify that some quantity is unchanged in natural processes. Examples: The total mass is unchanged in chemical reactions. Momentum is conserved in collisions.

Creationism. A pseudoscientific theory that poses as a theory of biological origins that rivals **evolution**. It ignores most scientific evidence and is motivated primarily by faith in the literal truth of the Bible.

Dendrites. Rootlike appendages of **neurons**, through which signals arrive from other neurons.

Determinism. The philosophical doctrine that every event, including human actions, necessarily results from causes.

Dualism (More precisely, **Cartesian** or mind-body dualism). The belief that human nature consists of two parts, a body and a spiritual entity such as a soul. The human abilities attributed to the soul often include life, consciousness, one's sense of morality, **free will**, and immortality.

Electric charge. A property of some **elementary particles** that causes them to exert forces on other charged particles. The natural unit of charge is denoted by e. An **electron** has a negative charge of magnitude e and a **proton** has a positive charge of magnitude e. **Neutrons** are uncharged (neutral). Atoms with equal numbers of electrons and protons are neutral, since the positive proton charges cancel the negative electron charges. *Static electricity* is due to a net charge on a body, due to an excess or lack of electrons. *Electrical currents* are usually produced by streams of electrons flowing through conductors.

Electromagnetic radiation. A form of energy that is emitted by charged **elementary particles** when they are accelerated. Radio waves, infrared radiation, light, ultraviolet radiation, X-rays, and **gamma rays** are all forms of electromagnetic radiation.

Electron. An extremely important negatively charged **elementary particle** that occupies most of the volume of atoms.

Elementary particle. Roughly equivalent to subatomic particles. Elementary particles are particles that are not known to be made of more basic particles. Since their discovery, some "elementary particles" have been found to be made of still more "elementary" **quarks** and **gluons**, but physicists have kept the earlier name.

Emergent property. A system's property revealed at a higher level, but not apparent at a more basic level. For example, a crystal can form from a large array of atoms, but direct observations of individual atoms do not reveal any properties of the crystal. Nevertheless, careful study of atoms can allow one to realize their potential for forming crystals, and even to predict the crystal's properties.

Endocrine glands. Organs that produce hormones, such as adrenaline, estrogen, and insulin, and release them directly into the bloodstream.

Enzyme. A biochemical that enhances or initiates specific chemical reactions in living cells. Like catalysts in chemistry, enzymes are not used up in the reaction process.

Eukaryote. Life-forms, distinguished from **prokaryotes**, by having cells with most of the genetic material in a **nucleus** surrounded by a membrane, and having various other structures within the cells. Most plants and animals are eukaryotes, except for bacteria and blue-green algae.

Evolution. Biological evolution is the process by which life has diversified from extremely simple beginnings. By **natural selection**, life-forms develop which are well adapted to their environments.

Final judgment. In Christian apocalyptic beliefs, the belief that Christ will return at the end of the world, that the dead will be raised, and that everyone will be judged to determine who will be sent to heaven or hell. Roughly similar ideas are present in Zoroastrianism, Judaism, and Islam.

Free will. Our feeling that we can make our own choices. Some people maintain that free will refers to one's power to make arbitrary choices that differ from the outcomes that would follow from natural causes.

Gamma ray. Very energetic **photons** with extremely short wavelengths. They are produced in decays of some **elementary particles** and atomic **nuclei**, and in interactions of energetic charged particles.

Gluon. A particle that binds **quarks** together inside **elementary particles**. Its name comes from "glue," due to its binding property.

Homunculus. A character in a naive picture of how consciousness "really happens." It is imagined that a small person (the homunculus) resides inside the brain, viewing a screen that displays what is happening in the outside world.

Hypothalamus. A pea-sized hormone-secreting region of the brain involved with the control of emotions and drives.

Innate value. As used here in connection with the formation of one's likes and dislikes, an instinctive appraisal of a situation as good or bad. For example, an infant's feeling that pain is bad.

Isotope. A form of a chemical element with **nuclei** containing a specific number of **neutrons**. Since the number of neutrons in the nuclei of an element may vary, many elements have a number of different isotopes.

Jesus movement. A group of very early Christians who produced the book of **Q**. They reported Jesus' sayings, but did not mention that he died for our sins or that he was resurrected.

Kinetic energy. Energy carried by matter due to its motion.

Life soul. In an ancient belief, a supernatural entity, the life soul, is needed to give life to an inherently lifeless body.

Lifetime (of **elementary particles**). The mean time interval between an elementary particle's creation and its decay into other types of particles. Some particles appear to have an infinite lifetime, while others live much less than a millionth of a second.

Limbic-brainstem system. Used by Gerald Edelman to refer to the combined **limbic** and **brainstem** systems. These parts of the brain control many processes automatically, outside our awareness. Examples are the control of heartbeat, body temperature, digestion, etc. This system is also deeply involved in generating our drives and emotions.

Limbic system. A name that refers to a number of brain areas near the bottom of the cerebrum, including the **thalamus**, **hypothalamus**, hippocampus, septum, and amygdala. Parts of this system are involved in producing emotions and memories.

Materialism. A philosophical view that all things that happen, including our thoughts and feelings, are due to matter and its interactions.

Mesons. Elementary particles made of a **quark**-antiquark pair.

Minor miracle. In one version of the **free-will** idea, more than one choice is possible when a person makes a decision, even for a fixed set of detailed circumstances. This implies that some **neurons** in one's brain can function either the way they should according to the laws of science, or in a different way, producing a different decision. As used here, a minor miracle happens when a decision is made that requires neurons to operate in a way that is normally impossible.

Motor nerve. Any nerve that controls movements by carrying signals from the brain to the body's muscles.

Mystery cults. Numerous religious cults that developed in ancient Greece and the Roman Empire, open only to those who were properly initiated into their secret rites and doctrines. To their members, they offered brotherhood or sisterhood, instruction in morality, assistance in meeting life's challenges, and hope for immortal life.

Natural selection. In biological **evolution**, the means by which species adapt to their environments. Basically, individuals with well-adapted genes have an advantage in surviving and reproducing, so the favored genes become more abundant in the general population. This allows the species, as a whole, to be well-adapted, even in changing environments.

Natural soul. As used here, a name for a set of capabilities which are entirely natural functions of the human nervous system. The natural soul includes all the abilities traditionally attributed to the supernatural soul, except immortality.

Neuron. A nerve cell, whether in nerves or in the brain.

Neurotransmitter. Certain biochemicals that carry neural signals from one nerve cell to another across a narrow gap that separates the cells.

Neutron. An **elementary particle** that carries no **electrical charge** and is found in atomic **nuclei**. It can persist indefinitely in a nucleus, but outside a nucleus it decays into other particles, with a mean **lifetime** of about fifteen minutes.

New System (NS). As used here in discussing the possible origin of our motives and feelings, the NS is the part of the brain that participates directly in conscious processes.

Nucleon. One of the **elementary particles** that makes up an atomic **nucleus**—a **proton** or a **neutron**.

Nucleus. In atomic physics, the nucleus is the central part of the atom, which accounts for almost the entire mass of the atom. It consists of **nucleons**. In biology, the nucleus is a part of the cell, enclosed in a separate membrane, which contains almost all of the cell's genetic material. It must be present if the cell is to grow and divide.

Objective Free Will (OFW). A form of the **free-will** belief in which one believes that, when one makes a decision, more than one choice was actually possible, with all the circumstances exactly as they were at the time of the decision. See **Minor miracle**.

Old System (OS). In this book's discussion of the origin of our motives and feelings, the OS is the part of the brain in which instinctive and unconscious processes occur.

Pattern recognition. The ability of organisms (or machines, if properly constructed) to detect the presence of certain visual shapes. Pattern recognition, for example, can enable animals to identify their enemies by observing them.

Photon. All electromagnetic radiation, such as radio waves, light, X-rays, and **gamma rays** consist of streams of **photons**, extremely important **elementary particles**.

Prefrontal cortex. A region in the front of the **cerebral cortex** that may be involved in planning, attention, **working memory**, and thoughts that trigger emotions.

Prokaryote. Relatively simple organisms, including bacteria and blue-green algae, which have a simpler cell structure than other life-forms, which are called **eukaryotes**.

Protein. Complex biochemical compounds that play a major role in the structure and function of animal and plant cells. DNA contains coded instructions for constructing an organism's proteins by linking large numbers of amino acids.

Proton. An **elementary particle** with a positive **electrical charge**. Along with **neutrons**, protons form atomic **nuclei**. Each chemical element has a unique number of protons in its nucleus.

Q. In New Testament studies, many scholars believe that Q and Mark were the two source documents for Matthew and Luke. Q may contain some of the oldest material in the New Testament and some of the most authentic teachings of Jesus.

Quantum effect. Any natural occurrence predicted by **quantum theory**, but contrary to **classical physics**. Quantum effects produce certain types of random behavior of atoms and **elementary particles**. They also allow many physical variables in microscopic systems to take on only certain discrete values.

Quantum theory. A theory of modern physics used primarily to describe the behavior of microscopic objects such as atoms and **elementary particles**. It successfully describes the microscopic world, although the observed behavior differs strongly from human intuition, which is based on our experiences in the (large-scale) everyday world.

Quark. All of the strongly interacting **elementary particles** are made of more basic particles called "quarks." **Nucleons** are made of three quarks, while **mesons** are made from a quark and an antiquark. Quarks are always found in combination with other quarks or antiquarks, never alone.

Radioactivity. The spontaneous emission of **gamma rays** or particles by atomic nuclei.

"Responsible." As used here, and always written in quotation marks, a special definition of responsibility that says we are not "responsible" for any of our actions that are ultimately and completely shaped by outside causes.

Resurrection. Returning to life by rising from the dead. In Christian doctrine, it can refer to the Resurrection of Jesus or to the general resurrection of all people at the time of the **final judgment**.

RNA (Ribonucleic acid). Complex biochemical compounds that are essential for the process of decoding DNA instructions and assembling **proteins**.

Spectrum. A distribution of **electromagnetic radiation** separated by wavelength. For example, a prism can separate sunlight into a spectrum of visible colors ranging from red to violet. The entire electromagnetic spectrum is much broader, ranging from radio waves to **gamma rays**.

Subatomic particle. See **Elementary particle**.

Subconscious process. A brain process that occurs without a person being aware of it. For example, breathing is controlled by the brain and is normally a subconscious process, although it can be brought under conscious control.

Subjective Free Will (SFW). Our feeling that we make our own choices and that they are meaningful. It may include a belief that the will is impossibly powerful. For example, one might believe that one could leap over the Grand Canyon by exercising enough willpower.

Synapse. A connection between **neurons**, consisting of a terminal on a branch of an **axon** from a signaling neuron; a narrow fluid gap, through which the signal is carried by a **neurotransmitter**; and the end of a **dendrite** (or sometimes the cell body) of the receiving neuron.

Temporal lobe. A region of the **cerebral cortex**, on the side of the head. This region is involved with triggering emotions, storing certain memories, recognizing objects, hearing, and, no doubt, many other functions.

Thalamocortical system. In the brain, the **thalamus**, the **cerebral cortex**, and the connections between them. It is involved in all aspects of mental activity, and is essential for consciousness.

Thalamus. An area located near the middle of the brain, between the two cerebral hemispheres. It is active during conscious activity, it controls the brain during sleep, and it may be involved in the process of attention.

Value signal. As used here, a neural signal in the brain that steers our decisions toward actions that tend to enhance our biological fitness. When we are conscious of these signals, they are our *motivations and feelings*.

Visual cortex. An area of the **cerebral cortex** near the back of the head, where visual information is mapped in an orderly fashion.

The images that we actually experience seem to be formed in the visual cortex, not in the eye's retina.

Working memory. The brain's temporary storage memory used in conscious mental processes. The contents of working memory are available for thought processes, planning actions, speech, introspection, and conversion to long-term memories.

INDEX